STICK AND RUDDER

STICK and RUDDER

An Explanation of the Art of Flying

WOLFGANG LANGEWIESCHE

Special Appendix on
THE DANGERS OF THE AIR
By LEIGHTON COLLINS

Illustrated by JO KOTULA

McGRAW-HILL BOOK COMPANY
New York . . . . *London*

STICK AND RUDDER

32 33 34 MAMM 7 6 5 4 3

ISBN: 07-036240-8

CONTENTS

PART I WINGS

PART II SOME AIR SENSE

PART III THE CONTROLS

PART IV THE BASIC MANEUVERS

PART V GETTING DOWN

STICK AND RUDDER

PART 1

WINGS

Get rid at the outset of the idea that the airplane is only an air-going sort of automobile. It isn't. It may sound like one and smell like one, and it may have been interior-decorated to look like one; but the difference is—it goes on wings.

And a wing is an odd thing, strangely behaved, hard to understand, tricky to handle. In many important respects, a wing's behavior is exactly contrary to common sense. On wings it is safe to be high, dangerous to be low; safe to go fast, dangerous to go slow. Generally speaking, if you want the airplane to go up, you point its nose up; but point its nose up a little too much, and you go down in a stall or a spin. In landing an airplane, to make it sink down on the runway and *stay* down, you move the controls much as for an extreme upward zoom. In the glide, if you want to descend more steeply, you point your airplane's nose down less steeply; if you want to descend less steeply, you point the airplane's nose down more steeply! And—most spectacular contrariness of all—in emergencies, when the airplane is sinking toward the ground in a "mush" or falling in a stall or a spin, and you are afraid of crashing into the ground, the only way to keep it from crashing is to point its nose down and dive at the ground, as if you wanted to crash!

It is largely this contrariness of the airplane that makes flying so difficult to learn. For flying *is* difficult to learn—let nobody tell you otherwise. The accident record proves it, and so does the number of men barred from flight training or eliminated from training for lack of aptitude. What makes flying so difficult is that the flier's instincts—that is, his most deeply established habits of mind and body—will tempt him to do exactly the wrong thing. ¹n learning other arts that are comparable to piloting—sailing, for instance—skills, ideas, habits must be developed where there were none before. In learning the art of piloting, much carefully learned behavior, many firmly held ideas must first be forgotten and cleared out of the way, must actually be reversed!

And it is largely because of this contrariness to common sense that the conventional airplane sometimes requires "nerve" from its pilot. Again, let no one tell you that this is not so. There are situations in flying when he who "ducks," he who flinches, is lost. The most important example is the recovery from a stall at low altitude—getting that stick forward and pointing the nose at the ground; that does require courage, and no two ways about it. But

3

there are many lesser examples as well. One of the government manuals on flying puts it that the pilot must learn not to give in to his instinct of self-preservation, but to substitute for it carefully trained reactions. That is only a very polite way of saying "guts."

From all this it might seem that learning to fly the conventional airplane must necessarily be mostly a matter of drill, like animal training, like making a dog not eat when he wants to eat, making him jump through a flaming hoop when he does not want to jump. And in fact, there is much of animal training in our flight training methods, at present, necessarily: for you simply cannot go against your common sense, against your most powerful instincts, except by drill, and more hard drill.

But another view of the problem is also possible. It may be that our common sense, our natural reactions mislead us simply because they are working on the basis of wrong ideas in our minds concerning the wing and how it really flies, the controls and what they really do. The airplane, after all, is a machine; it obeys ordinary physical law. Its behavior cannot really be contrary to common sense. The wing is simple—just as simple as man's other really great inventions, the wheel, the boat, the lever; but, unlike those other inventions, the wing is new and not yet familiar to us of this generation.

Perhaps what happens when the beginner reacts wrongly in an airplane is similar to what happened in the early days of the automobile, when a man trying to stop in an emergency would pull back on the wheel as if he had reins in his hands and would even yell "Whoa." There was nothing really wrong with his reactions, with his intentions; the only thing wrong was the image in his head that made him see the automobile as a sort of mechanized horse, to be controlled as horses are controlled. Had he clearly seen in his mind's eye the mechanical arrangement we take for granted now—the clutch that can disconnect the motor, the brakes that can clamp down on the wheels; had he clearly appreciated that the thing was a machine and had no soul at all, not even a horse's soul, and that thus there was no use in speaking to it—he would then have done the right thing without difficulty. It may be that, if we could only understand the wing clearly enough, see its working vividly enough, it would no longer seem to behave contrary to common sense; we should then expect it to behave as it does behave. We could then simply follow our impulses and "instincts." Flying is done largely with one's imagination! If one's images of the airplane are correct, one's behavior in the airplane will quite naturally and effortlessly also be correct.

The effort to understand how an airplane flies is sometimes called "Theory of Flight." Under that name, it has a bad reputation with pilots. Most pilots think that theory is useless, that practice is what does it. Yet you can't help having a theory: whatever you do, from peeling potatoes to flying airplanes, you go on the basis of some mental image of what's what—and that's all "theory" amounts to. And if your ideas of what's what are correct, you will do it well.

What is wrong with "Theory of Flight," from the pilot's point of view, is not that it is theory. What's wrong is that it is the theory of the wrong thing—it usually becomes a theory of building the airplane rather than of flying it. It goes deeply—much too deeply for a pilot's needs—into problems of aerodynamics; it even gives the pilot a formula by which to calculate his lift! But it neglects those phases of flight that interest the pilot most. It often fails to show the pilot the most important fact in the art of piloting—the Angle of Attack, and how it changes in flight. And it usually fails to give him a clear understanding of the various flight conditions in which an airplane can proceed, from fast flight to mush and stall. This whole book, and especially its first chapters, are an attempt to refocus "Theory of Flight," away from things that the pilot does not need to know about, and upon the things that actually puzzle him while he flies.

Chapter 1

HOW A WING IS FLOWN

AT THIS very moment, thousands of men, trying to learn to fly, are wasting tens of thousands of air hours simply because they don't really understand how an airplane flies; because they don't see the one fact that explains just about every single thing they are doing; because they lack the one key that with one click unlocks most of the secrets of the art of flying.

In the textbooks, this thing is discussed under the name of Angle of Attack. The story of the Angle of Attack is in a way *the* theory of flight: If you had only 2 hours in which to explain the airplane to a student pilot, this is what you would have to explain. It is almost literally all there is to flight. It explains all about the climb, the glide, and level flight; much about the turn; practically all about the ordinary stall, the power stall, the spin. It takes the puzzlement out of such maneuvers as the nose-high power approach; it is *the* story of the landing. No maneuver can be fully understood unless you understand this one thing. You may then still not be able to fly well; you may still be clumsy at moving the stick and rudder perfectly together. Your eyes and ears and feel may still be a little dull; but you will understand flying and not be puzzled; you will be able to figure out what you ought to do; you will be able to analyze your own mistakes; and you will get by.

To understand this thing, however, requires real mental effort. There is only one easy way to understand unknown things—comparison with known things. You can understand the effects of controllable pitch on a propeller by comparison with the gears of an automobile; a propeller itself, if you like, by comparison with a screw; a rudder by comparison with a ship's rudder; but the wing is the one thing about the airplane that is new and is peculiar to airplanes alone. And thus the Angle of Attack has no similes in our life on the ground.

First of all, what Angle of Attack is *not;* for the word is often misused. The Angle of Attack is *not* the angle which your nose makes with the horizon; it is *not* the angle at which your ship points up or down. Pilots sometimes use the word in that sense. The word one ought to use for this is "attitude": a nose-high attitude, a level attitude, a nose-down attitude. The distinction is important, for an airplane may be in a quite nose-low attitude and yet at very high Angle of Attack, for example, during a spin. Again, an airplane may be pointing almost straight up and yet be at fairly low Angle of Attack, for example, during a zoom.

Nor is the Angle of Attack the angle at which your wings are set with reference to the lengthwise axis of your airplane. That angle is called the *angle of incidence;* * it does not concern the pilot much, for it is built into the airplane by the designer, and the pilot can do nothing about it. Moreover, in most airplanes, the angle of incidence is negligibly small—one or two degrees or even zero. Hence, from now on in this discussion, the angle of incidence will generally be disregarded; it will be assumed that the airplane is built with zero angle of incidence.

All that is what Angle of Attack is *not*. Here is what it *is;* and this is the idea you will want to mull over, to draw on paper, to act out with your hands, to try out in flight—until it finally becomes completely clear and familiar: *the Angle of Attack is the angle at which the wing meets the air.*

FORGET BERNOULLI'S THEOREM

The angle at which the wing meets the air; what does it mean? To understand this, you have to go back to a simple idea of how a wing really manages to fly, how lift is developed. When you studied theory of flight in ground school, you were probably taught a good deal of fancy stuff concerning an airplane's wing and just how it creates lift. As a practical pilot you may forget much of it. Perhaps you remember *Bernoulli's Theorem:* how the air, in shooting around the long way over the top of the wing, has to speed up, and how in speeding up it drops some of its pressure, and how it hence exerts a suction on the top surface of the wing. Forget it. In the first place, Bernoulli's Theorem does not really explain--the explanation is more puzzling than the puzzle! In the second place, Bernoulli's Theorem doesn't

* In British usage, angle of incidence means what Angle of Attack means in American usage.

help you the least bit in flying. While it is no doubt true, it usually merely serves to obscure to the pilot certain simpler, much more important, much more helpful facts.

Perhaps you also remember that rather highbrow concept of *circulation*—how (in a manner of speaking at least) the air flows

The airplane keeps itself up by beating the air down. If one could lay a smoke cloud into an airplane's path . . .

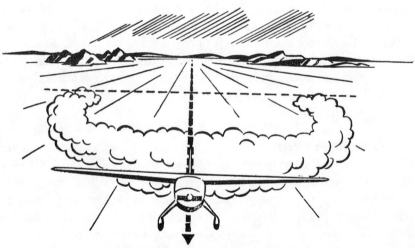

. . . the airplane, in flying through it, would squash it down.

around the wing, forward under the wing's bottom surface, and backward over its top surface; and how in doing so, it creates lift. Forget that, too. It, too, is no doubt true, though of course an abstraction; and it is no doubt useful knowledge for engineers. For a pilot it, too, is useless knowledge; it, too, can be actually harmful if it is allowed to obscure the simpler, more fundamental facts of flight.

The main fact of all heavier-than-air flight is this: *the wing keeps the airplane up by pushing the air down.*

It shoves the air down with its bottom surface, and it pulls the air down with its top surface; the latter action is the more important. But the really important thing to understand is that the wing, in whatever fashion, makes the air go *down.* In exerting a downward force upon the air, the wing receives an upward counterforce—by the same principle, known as *Newton's law of action and reaction,* which makes a gun recoil as it shoves the bullet out forward; and which makes the nozzle of a fire hose press backward heavily against the fireman as it shoots out a stream of water forward. Air is heavy; sea-level air weighs about 2 pounds per cubic yard; thus, as your wings give a downward push to cubic yard after cubic yard of that heavy stuff, they get upward reactions that are equally hefty.

That's what keeps an airplane up. Newton's law says that, if the wing pushes the air down, the air must push the wing up. It also puts the same thing the other way 'round: if the wing is to hold the airplane up in the fluid, ever-yielding air, it can do so only by pushing the air down. All the fancy physics of Bernoulli's Theorem, all the highbrow math of the circulation theory, all the diagrams showing the airflow on a wing—all that is only an elaboration and more detailed description of just how Newton's law fulfills itself—for instance, the rather interesting but (for the pilot) really quite useless observation that the wing does most of its downwashing work by suction, with its top surface. Trying to understand the piloting of airplanes by concentrating on Bernoulli and Prandtl is like trying to catch on to tennis by studying just exactly how the rubber molecules behave in a tennis ball when the ball hits the court and just exactly how the catgut behaves in the racket when the ball strikes: instead of simply observing that it bounces!

AN INCLINED PLANE

Thus, if you will forget some of this excessive erudition, a wing becomes much easier to understand; it is in the last analysis nothing but an air deflector. It is an inclined plane, cleverly curved, to be sure, and elaborately streamlined, but still essentially an inclined plane. That's, after all, why that whole fascinating contraption of ours is called an air-*plane.*

It is the plane part of the airplane we have to understand.

This plane is *inclined* so that as it moves through the air, it will meet the air at an angle and thus shove it downward, in somewhat the same way that the inclined plane of a snowplow, in moving forward against the snow, shoves the snow to the side. And the angle by which it is inclined, the angle at which it meets the air, is for every pilot the most important thing in flying; for *that* is the Angle of Attack.

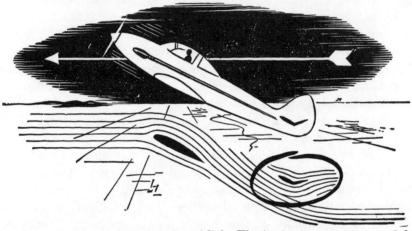

Slow flight. White arrow shows direction of flight. The Angle of Attack is large and the downwash is sharp. Upward-deflected flippers (pilot holds stick back) keep ship at large Angle of Attack.

FLY AND SEE

To demonstrate the Angle of Attack to yourself, it may be best to get into the airplane and fly out a simple flight problem. Suppose that for a couple of minutes you fly straight and level, carefully trimming the ship at cruising power, say 2,100 r.p.m. After everything has steadied down, cut back your power to about 1,400 r.p.m. And from then on, do whatever is necessary to maintain your altitude.

Of 100 student pilots to whom you might propose this idea, about 80 would say that it cannot be done! They would say the airplane will of course simply stall—which shows that something is lacking in our flight training, that we allow students to handle a machine of which they don't understand the first thing. For actually, level flying with much reduced power is easy to do. With the front seat empty, most training airplanes will maintain level flight comfortably at about two-thirds the usual cruising r.p.m.! All you have to do is hold your nose up by continuous back pressure on the stick. Then you simply proceed at very slow speed, nose-high. Your ship *points* upward as if it wanted to climb, but its actual flight *path* remains level.

It is a most interesting condition of flight. For it is an exaggerated demonstration of that most important of all flying facts—Angle of Attack. The Angle of Attack can also be defined as the difference between where the airplane points and where (in the up-and-down sense) it goes. Pilots call this flight condition "mushing." An engineer would call it simply "flight at high angle of attack." At any rate, a man isn't much of a pilot until he has realized (by feel and gradual experience—or else by reasoning and experiment) that it is perfectly possible and proper for the airplane to proceed in that fashion. True enough, the ship is close to the stall; but it is not stalled, and it can fly in this fashion forever, except that in some types the engine may gradually overheat.

MUSHING?

In fact, one might almost say that this is the normal way for the airplane to fly. It is only a dramatized and exaggerated demonstration of how it really always flies. An airplane always flies nose-high, *pointing* up a little higher than it actually *goes*. At least, an airplane's wing does; if it didn't, it would have no Angle of Attack. If it had no Angle of Attack, it would not wash the air down; and if it did not wash the air down, there would be no lift. In fast cruising, the Angle of Attack is very small. In slow flight, this small Angle of Attack does not provide enough lift to hold the airplane up. This is because, at the slower speed, the wing catches fewer pounds of air per minute; hence, in order to keep the lift strong enough to hold the airplane up, the air must be deflected more sharply downward; to accomplish this, the wing's angle of attack must be increased; the pilot does this by holding the stick farther back. But still there is no real difference between cruising flight and slow, mushing flight. The only difference is this: in ordinary flight, the Angle of Attack is so small that the student pilot does not realize its existence; in very slow flight, the Angle of Attack is so large that even the student pilot suddenly realizes what goes on. He notices that the airplane does not go where it is pointed, that there is an admixture of *sink* in its motion. This is likely to disconcert him:

"Good heavens," he thinks, "I'm mushing! What do I do now?" Actually there is nothing abnormal, nothing wrong. He has merely realized for the first time the ordinary facts of flight.

You will also notice that, in flight at such high Angle of Attack, the control feel is rather different from that of cruising flight. The ailerons feel soft; and, unless the stabilizer is trimmed well back, there is need for continual strong back pressure on the stick to hold the airplane at that high Angle of Attack. But then, odd control feel does not prove that anything is wrong. The controls feel different for each different Angle of Attack (or, what is the same thing, for each different speed). They feel quite odd, for example, also in a fast dive. Finally, you will probably realize that in this very slow flight your margin of safety over the stall is much reduced; but as long as you do realize it and act accordingly, there's nothing wrong with that. In short, then, slow nose-high level flight is a perfectly normal, healthy, and steady condition.

AN ASIDE TO FLIGHT INSTRUCTORS

It would probably be good for the beginning student pilot to get some experience in this kind of flying quite early—even before starting in on landings. It might not pay off immediately in actual flying skill, but it would certainly pay off by giving the student a better understanding of the whole process of flight. In the course of ordinary flight training, the student experiences very slow flight rarely, and only for moments at a time. He gets it during landings, during stall practice, and also perhaps sometimes on a take-off when he has lifted the ship off prematurely, nose-high, at too low a speed. In all these cases, he does not stay in this condition; on the contrary, in all these cases, he tries to get out of it, either trying to stall the airplane, as in landings and in stall practice, or else trying to recover "normal" flight, as after a premature nose-high take-off. Hence he may come to think that an airplane cannot maintain this flight condition steadily; that if it were held thus for more than a few seconds, it would stall. And if he believes that, he misses the main point of the whole art of flying.

If a student were given 5 minutes of very slow level flight a few times before starting him out on landings, it might even pay out directly, in actual skill. Here is why: The time a student actually spends on landing during his whole elementary flight training amounts only to something like 10 minutes!* This means, of course, the time spent on the actual landing process—not on rolling on the ground and take-offs and circuits of the airport and approach glides, but on the actual landing process—leveling off and "holding off." No wonder landings seem so difficult!

The entire process of the three-point landing is essentially one of slower and still slower flight, flight at higher and still higher Angle of Attack; all the time exercising very accurate control over the flight path. Yet, at the time when the student first tries landings, he has practically no experience at all in flight at medium or high Angle of Attack. Thus he must learn three things all at once, all of them brand new—first, that the airplane *can* fly in this fashion at all—second, how it responds to the controls in this kind of flight—third, how to judge its flight path so that contact with the ground will be smooth. He has to get acquainted with flight at high Angle of Attack under the most difficult condition, that is, near the ground. And he has to get acquainted with the ground under the most difficult condition, that is, while flying at high Angle of Attack.

True, the instructor usually prepares the student by giving him some power-off stalls at altitude before giving him landings. This is no doubt better than no preparation at all; but it still does not give the student much chance to become really at home in slow, nose-high flight. For in stall practice, too, the ship goes through the whole range of Angles of Attack rather fast. Sometimes such practice may even reinforce the idea that, whenever the nose is high, a stall will inevitably result within a few seconds.

TOWARD NORMAL FLIGHT

At any rate, if the airplane seems as yet a little new and strange to you, it is worth your while to take your ship up once to a safe altitude and fly it a few minutes in this fashion, maintaining altitude in nose-high flight with very little power. For what you observe there is the very heart of the matter.

Suppose now you continue your flight experiment. Open your throttle a bit wider and then do whatever is necessary to maintain a strictly level flight path. Can it be done? Obviously it can. The ship will climb; you will relax some of your back pressure against the stick,

* An observation due to Dr. Dean R. Brimhall.

and the ship will then no longer climb, but the nose will come down lower, and the ship will speed up.

Flight now, at perhaps 1,800 r.p.m., is only a less dramatic version of the same process that you observed when you were mushing along at 1,400 r.p.m. Because the wings now meet the air at higher speed, they need not turn it downward quite so sharply and hence need no longer meet it at quite so large an Angle of Attack. But of course they still have an Angle of Attack, and they still wash the air down. It is still true that the airplane keeps itself up by pushing the air down.

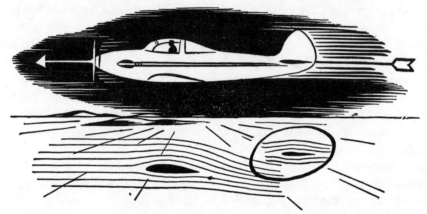

Fast flight. Angle of Attack is almost invisibly small, downwash slight. Flippers neutral. This is normal cruising condition.

CRUISING

Now, as a third step in your experiment, suppose you advance your throttle to regular cruising power, assumed in this instance to be 2,100 r.p.m. Again, in order to hold level flight and keep from climbing, you will have to hold the nose down farther. To accomplish this, you will now have to release all back pressure on the stick. And there you are, back in level cruising flight, the flight condition that seems normal because we do most of our flying in it.

Now this condition, cruising flight: is it essentially at all different from that nose-high "mushing" sort of flight that you had earlier? It may seem so. For now, when the airplane is flying level it also *points* level; in other words the airplane points where it is going; or, in still other words, the airplane now actually goes (in the up and down sense) where its nose is pointing. It might seem as if there were now no Angle of Attack—as if it now maintained flight by some principle other than that which keeps it up in "mushing" flight.

But there is no other principle. There is an Angle of Attack. There must be. If there were not, no air would be pushed down, and short of having old Bernoulli himself hold it up with a hook, there simply is no way for a ton or two of machinery to stay up in the thin air. It has no way to keep itself up except by continually beating the air down. The difference is again only one of degree, not of kind. It is only that now, with the speed so brisk, an exceedingly small Angle of Attack is

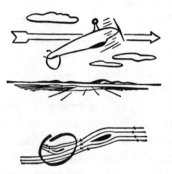

You fly upside down by the same principle by which you fly right side up: the wing meets the air at an Angle of Attack and washes it down. Only difference: wing section, used upside down, is inefficient as a down deflector. Hence a large Angle of Attack is needed to produce enough downwash and thus enough lift. Note "up"-ward deflected flipper: pilot holds stick forward to keep Angle of Attack large.

enough to produce the necessary lift—the wing meets the air at perhaps 1 or 2 degrees Angle of Attack—an angle too small to be noticeable to the eye. The fuselage points level in cruising flight—simply because the designer joined it to the wings at such an angle (the angle of incidence) that, with the wings in level flight at cruising Angle of Attack, it *would* point level!

WIDE OPEN

Next, suppose you pursue your experiment one further step: you open your throttle all the way and *still* do whatever is necessary to maintain a level flight path. Again, of course, you have to put your nose down farther to keep from climbing, and you will have to hold it down by continual forward pressure on the stick. Again this results in an increase in speed. Presently you find yourself hurrying along with the nose pointing definitely down, as if you were in a shallow dive, and yet holding your altitude! Apparently the Angle of Attack is now negative!

"Now," one might say, "what is holding her up? Doesn't this prove that there is something else to a wing's lift—some principle other than Angle of Attack and the downwashing of air?"

The usual ground-school answer is simply that "it so happens" that a wing will develop lift even while at negative Angle of Attack, and Bernoulli is once more cited in "explanation." The way we reckon Angle of Attack at present, this is a true enough statement. But it sounds absurd; and it is harmful if it suggests that the wing is developing lift without washing the air down, which is an impossibility. The whole problem, however, is merely one of vocabulary.

It is customary to reckon Angle of Attack as the angle that the *chord* of the wing makes with the oncoming air. Chord, in other words, is the reference line. But the chord is not what really counts in a wing. It is used by the practical engineers only for the sake of convenience— because it is easily measured. The line that really counts is the *no-lift line;* and the more highbrow research men all talk in terms of "absolute" Angle of Attack—the angle that the no-lift line makes with the oncoming air. And the no-lift line of a wing is still at positive Angle of Attack, even in full-power very fast flight; the wing is still an inclined plane, meeting the air at an Angle of Attack; and it is still pushing the air down.

IT ONLY LOOKS THAT WAY

Perhaps the following train of reasoning, though quite unscientific in detail, will help you to keep your ideas concerning lift simple and clear. Say that the wing is basically simply a plane, set at a slight inclination (the Angle of Attack) so as to wash the air down. This inclined plane is shown on page 17. But it was early found that the drag, lifting, and stalling characteristics of such an inclined plane can be improved by surrounding it with a curving, streamlined housing; hence our present wing "sections." The actual wing of an airplane is therefore not simply an inclined plane; it is a curved body *containing* an inclined plane. And this basic inclined plane, which is (in an imaginary fashion) *contained* of every wing, is the no-lift line when you look at the wing in cross section. The curved streamlining is arranged around this plane, as shown on page 17; that is, it is set around it in an unsymmetrical fashion. Hence, in very fast flight, when the inclined plane is working at very small Angle of Attack, the wing looks, because of its streamlined housing, as if it were at no Angle of Attack, or even at negative angle. But it only looks that way; essentially, it still meets the air at an Angle of Attack; and it still makes lift simply by washing the air down.

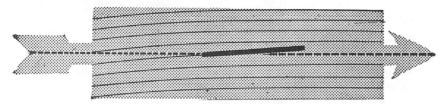

In basic idea, a wing is an inclined plane, set into a wind so as to deflect the air downward.

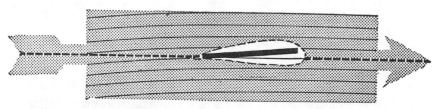

For greater efficiency, this basic inclined plane is enclosed by a curved outer shape. This sometimes makes it appear as if the wing were at zero or negative Angle of Attack when its real ("absolute") Angle of Attack is still positive. "Chord," is the line from the frontmost to the rearmost point of the wing section—thin solid line in this picture. In aviation *practice*, Angle of Attack is reckoned as the angle between the chord and the Relative Wind. In that sense, it is true that a wing can develop lift at zero or even at negative Angle of Attack. In aerodynamic *theory* Angle of Attack is reckoned as the angle between the Relative Wind and the "no-lift line"—the no-lift line being really our basic inclined plane, seen in cross section. And in that sense, it is true that a wing cannot develop lift unless it has an Angle of Attack.

THE LESS SPEED, THE MORE ANGLE OF ATTACK

Actually, then, the process of flight is the same, whether you are just mushing along in slow flight or cruising or going at top speed; the wing meets the air at an Angle of Attack, kicks the air down, and thereby kicks itself up.

For every speed, there is one Angle of Attack that will produce just enough lift to hold your ship up. The more speed you have, the less Angle of Attack you need; the less speed you have, the more Angle of Attack you need. In the fancy language of the engineers, the words *speed* and *Angle of Attack* are therefore used almost interchangeably. Instead of saying the ship was flying fast, the engineer prefers to say, "It was at low Angle of Attack." But weight, too, has something to do with Angle of Attack; if you load more weight into the airplane, then for any given speed it will have to have more Angle of Attack in order to sustain itself in the air. That is why heavily overloaded airplanes, setting out on extreme long-distance flight, often have to "mush" along, nose-high, for the first few hours even though the engines are perhaps working nearly wide open; until enough fuel has been burned up to lighten the ship, so it can fly at more normal Angle of Attack.

WHY DO WE STALL?

Suppose now you make a final experiment at a safe altitude. Set your throttle for about 800 r.p.m., and *now* try to maintain your altitude.

You know what will happen; it can't be done. In the attempt to maintain altitude, you will pull the stick back; the airplane will slow up; you will have to pull the stick back still farther, and the nose still higher—and eventually, you will stall. What too many pilots do not understand is just why the stall really occurs and how it is tied up with this whole matter of Angle of Attack. For the fact is that Angle of Attack, which is the key to so many things in flying, is the key also to the puzzle of the stall.

IT'S NOT BECAUSE OF LACK OF SPEED

First of all, let's check on some ideas that are wrong. Many students think that the direct cause of every stall is lack of speed. "The air flow over the top of the wing is not brisk enough and there isn't enough lift." That isn't it. A stall is not directly caused by lack of speed.

It is possible to stall an airplane at speeds very much higher than usual by loading the airplane up excessively with centrifugal force. In a 60-degree banked turn, for instance, your stalling speed will be nearly one and a half times as high as it is in normal straight flight. Somewhat the same will be true during a sharp pull-out from a dive. It is possible to stall your airplane at any speed, even at top speed, simply by pulling the stick back far enough abruptly enough!

Again, it is quite possible for a wing to develop lift when at very slow speed—'way below the usual stalling speed. For example, imagine that you are taxiing a light plane at 20 m.p.h. Say that you are taxiing it with the tail just slightly off the ground. In this condition, its wings will develop considerable lift. It won't be quite enough to sustain the whole weight of the airplane, and hence you won't leave the ground. But it will be enough to take most of the weight off the wheels, and if you are taxiing on sand or mud, that will be quite noticeable and helpful. In short, lack of speed is not the direct cause of a stall. Plenty of speed is not necessarily a protection from the stall.

IT'S NOT BECAUSE THE NOSE IS TOO HIGH

Some students think that an airplane stalls because the nose is too high. This is true, too, after a fashion; the question is what constitutes "too high." Under some conditions, an airplane can point straight up into the sky and yet not stall; for example, when it is being pulled up into a loop. Under other conditions an airplane can stall with its nose well below the horizon; for example, during a steep turn with power off. Whether or not the nose is too high depends on the amount of power the airplane has and on the type of maneuvering you do with it; in short, it depends on so many factors that it becomes meaningless to say that an airplane stalls because its nose is too high.

THE AIR CAN'T TAKE THE CURVE

The direct and immediate cause of any stall is always only one thing: *excessive Angle of Attack*—"excessive" meaning, for most wings, greater than 18 degrees. Whenever a wing meets the air at a proper, moderate Angle of Attack it will act as a downward deflector of air, will wash the air down, and hence will experience an upward lift. The lift may not be enough to maintain the airplane in flight, but the wing will not be stalled. But whenever a wing meets the air at too large an Angle of Attack, and tries to wash it down too sharply, the air fails to take the downward curve. The air flow over the top of the wing burbles and breaks away from the guidance of the wing's curved top surface. The wing is then no longer an efficient downward deflector of air. It still causes a large commotion in the air, but this commotion then consists largely of useless turbulence; it contains very little downwash. In causing this commotion, the wing therefore experiences much drag but very little lift.

That's what a stall is: the failure of the air to take the downward curve. And that's how a stall is caused: the excessive demands made on the air by a wing which meets it at too large an Angle of Attack.

True, lack of speed is indeed the most frequent reason why a pilot forces his wings to excessive Angle of Attack and *thus* stalls his airplane. But it is not the only cause. The wing can be brought to excessive Angle of Attack quite independently of speed. As will be explained later in this book, the control that determines Angle of Attack is the elevator. Thus, simply by pulling the stick back far enough, the pilot can stall his airplane at any speed. The classic example of this is the snap roll. At twice his ordinary stalling speed, the pilot pulls the stick back rather sharply and far. This causes the airplane to rear up while at the same time continuing, because of its inertia, in more or less the original line of motion. Thus the wings meet the air at an excessively large Angle of Attack, and they stall—even though the speed is high. This kind of stall is sometimes called a *snap stall* and can occur, regardless of speed, whenever the stick is brought back too abruptly and too far. In order to execute a snap roll, the pilot then kicks rudder, thus converting the stall into a spin; the snap roll is nothing but a horizontal power spin. This is described here not in order to explain the snap roll, much less to give a recipe for how to do one. On the contrary, the snap roll is used here to explain how an airplane stalls: the important thing to realize clearly is that the stall is the direct and invariable result of trying to fly the airplane at too large an Angle of Attack.

Now that this is clearly understood, we can go back to our flight experiment—the last phase of it, when the pilot is trying to maintain altitude on about 800 r.p.m., and stalls in the attempt. This experiment shows just why it is that every airplane has a certain speed below which it simply can't be flown.

When a pilot stalls an airplane in straight flight by trying to fly too slowly, here is what happens: as he flies more slowly he must increase his Angle of Attack in order to maintain enough lift to keep the ship in air. He does this by holding the stick farther back. The slower his flight the larger is the Angle of Attack he needs. Finally some speed will be found that is so slow that even a very high Angle of Attack just barely produces enough lift to keep the ship sustained. Then, suppose the pilot slows the ship up still a little more; in his attempt to keep the ship flying, he then increases his Angle of Attack still a little more, and he thereby exceeds the critical angle beyond which his wing cannot work; the wing stalls.

HOW BLOWS THE RELATIVE WIND?

But now for a more realistic picture of what really goes on when we maneuver. That first flight experiment was carefully set up so it would not be confusing. The flight path was at all times strictly horizontal; hence the air was flowing against the wing always strictly horizontally, from straight ahead; hence you could always *see* your Angle of Attack simply by looking out at the wings, marking the angle at which they held themselves with the horizon: how nose-high or nose-low they were.

In practical flying that is exactly what does *not* happen: our flight path is *not* necessarily level; even though the nose may be pointing level, the airplane itself may be sinking through or may be ballooning. Hence the air flows at the airplane *not* necessarily horizontally from straight ahead. It may, for instance, flow at the airplane upward, from ahead and below. Hence the Angle of Attack cannot be seen simply by looking out the window; in fact, *it cannot be seen* at all! For remember, Angle of Attack is the angle at which the wing meets the *air*—and we can't see air. That is perhaps largely why flying is so much of an art. In baseball the batter keeps his eye on the ball that he is going to hit. Flying is the art of batting the air down with our wings; but in flying, our trouble is that we can't see the air; hence we often fail to hit it right, and hence so many of us break our necks.

You might say (and many pilots do say), "If it is all so hard to understand, why bother? If this mysterious thing is so hard to see, why, let's not try to fly by it; let's fix our attention on something else, on something we *can* see or hear or feel."

The answer is that there *are* no other things we can go by; none, that is, of equal importance. If there were, flying would be much simpler. If you want to understand flight, you have to understand the Angle of Attack. And if you want to understand the Angle of Attack, then you have to understand just where, under the various conditions of an airplane's maneuvering, the air comes from which the wing is meeting.

THE WIND OF FLIGHT

This is usually discussed under the paragraph title The Relative Wind. This is a rather fancy word for a quite simple thing. Imagine a bicyclist saying, "The weather is hot and the day is calm but my relative wind keeps me cool." He means simply this: as long as he keeps moving, he gets "a sort of" breeze, produced by his own motion. That breeze is not the air's motion past him, but his own motion through the air: his *relative* wind. This "wind" blows at him only while he keeps going. If he stops, this "wind" also stops. Perhaps it should be called

The Relative Wind is the "breeze" made by your own motion relative to the air. The two skaters, moving in opposite directions, feel opposite Relative Winds.

The flag on the pole flutters in a real wind. The flag on the car flutters in the car's Relative Wind—produced by the car's motion.

the *wind of flight;* perhaps it would be best to avoid the word *wind* altogether, for of course it is not really a wind. The air stays still; it is the airplane that moves. Saying that there is a wind blowing against the wings of the airplane is a figure of speech, used just for convenience. It is like saying that the telegraph poles rush by the car, and the road rushes through underneath it. The air doesn't really rush against the airplane; the airplane rushes against the air. But the effects are the same either way—there is "a sort of" a wind.

This so-called "Relative Wind," this onrush of air, this "wind of flight" always comes at the airplane from the direction toward which the airplane is moving. If the airplane flies level, the Relative Wind blows back at it level. If the airplane slides sideways through the air in a sideslip to the left, this wind of flight blows back at the airplane's left side. In a glide, when the airplane moves downward along a sloping flight path, the wind of flight comes blowing up that slope. In a vertical dive, the wind of flight blows at the airplane straight up! In a steep climb, when the airplane climbs up a sloping flight path, the wind of flight blows down the same slope! In a spin, when the airplane is going down, the relative wind blows up; but, since in a spin the airplane goes down twisting, the resulting wind of flight blows up against it twisting!

When a man first starts flying, he judges mostly by "mechanical" clues; how the nose points above or below the horizon, for example;

A Relative Wind indicator. The masthead pennant on the sailboat—admittedly pictured slightly incorrectly—is nothing but a Relative Wind indicator. Similar pennant could be carried by airplane. In level cruising flight it would stream as shown on higher airplane. In a glide it would stream as shown on lower airplane.

how his throttle is set; what his instruments show. That is all right to begin with; but those things are not really the most important ones for flight. They are things of the ground, not things of the air. Then, as a man goes on and flies more, he discovers skid and slip, and develops a sense for his air speed. That's much more to the point. But it still is not the most important thing. The factor that has most to do with keeping an airplane up or making it drop, rendering it obedient to the controls or rendering it uncontrollable is the *direction* of the Relative Wind: the angle from which the air is rushing against the wing. When an experienced pilot flies an airplane, when he puts it through climbs and glides, turns, stalls and spins, take-offs and landings, he is asking himself all the time, "Where, at this moment, is the air coming from? And at what angle are my wings meeting it?"

Most pilots would of course deny that they ever think such a thing. And it is a fact that many excellent pilots don't even know the meaning of the words Angle of Attack and Relative Wind. But you may bet that, subconsciously, most good pilots' minds do function somewhat that way; though it is "feel" rather than deliberate reasoning. A good pilot's "feel" for his airplane, his almost instinctive ability to handle it right, is in the last analysis nothing but continual awareness of this most important of all flying facts—the Angle of Attack, which can also be defined as the angle at which the wing meets the Relative Wind.

Just what he knows about it, and just how he "feels" it, will now be explained.

Chapter 2

THE AIRPLANE'S GAITS

YOU know how a horse uses a different gait at different times: it may walk, or trot, or gallop, and in addition some of them use certain fancy gaits. An airplane, too, has distinct gaits; it uses sharply different modes of motion at different times—some of them obvious, some of them fancy. Consider the descent alone: an airplane may descend in a dive, or in a Normal Glide, or in a Mushing Glide, or in a Power Descent: four modes of getting down! All together, one can distinguish eight such gaits.

A student flier should understand those eight gaits before he ever gets off the ground. If he can gain a clear idea of how flight path and attitude, power and speed, Angle of Attack and Relative Wind shape up in each of these flight conditions, he will find relief from that frustrating sensation which usually marks one's early flight training: the vague feeling that he is missing the point of what's going on. What's more, once he understands those eight characteristic flight conditions, it will be rather easy for him to produce them in flight. And if he can produce them in flight, he is more than halfway a pilot!

CRUISING FLIGHT AND ECONOMY FLIGHT

Two of these gaits have been discussed and illustrated in Chapter 1: level flight at high Angle of Attack, and level flight at low, almost invisible, Angle of Attack. Level flight at low Angle of Attack is the airplane's normal gait; that's how it cruises. Its whole design aims primarily at efficiency in cruising flight.

Level flight at high Angle of Attack is less familiar, because it is of no special use to a beginner's flying. But in high-grade practical flying, it is quite important. It might be called "Economy Flight." As will be shown later in this book, an airplane will fly the most *miles* per gallon of fuel when flown rather slowly and nose-high; and it will keep flying the most *minutes* per gallon of fuel if flown very slowly, very nose-high. Hence most long-distance flights, including many bombing ✈ aids, are

A Dive.

made at this gait. Furthermore, this is the gait which the airplane assumes willy-nilly when flown at high altitude. Our final chapter will show that the much-discussed advantages of "stratosphere" flying are intimately tied up with the economy of this slow, nose-high gait.

THE DIVE

Next, consider the gliding gaits—that is, those flight conditions in which the airplane proceeds without help from its power plant. Of these, the easiest to understand is the dive.

The airplane is nosed steeply down. Instead of being pulled along by its engine, it is being pulled along by gravity. Apart from this one fact, the whole flight condition is quite similar to cruising flight: the airplane moves fast, and hence flies at low Angle of Attack. This means that it actually goes just about where its nose points—at least in a moderate dive. In cruising flight, it points and goes level. In a dive, it points and goes down.

An interesting form of the dive is the *vertical* dive—vertical in that the airplane *points* straight down. There is then almost no upward direction to the wings' lift. Instead, the wings tend to push the airplane along horizontally, parallel to the ground. And since this wing force is not met by any counterforce (as in normal flight, when it is an *upward* push, it is countered by the downward force of the airplane's weight) the airplane though *pointing* straight down does not actually *go* straight down: it travels quite a distance horizontally while it dives.

Vertical dive. Wings develop lift which pushes as shown by arrows. Airplane *points* straight down but flight *path* is not straight down.

No-lift dive. Airplane slightly inverted wings at zero (absolute) Angle of Attack. No lift. Flight path straight down.

The same effect is noticeable even in less steep dives: you pick out a certain field somewhere ahead of you and below you, and dive at it, pointing your nose straight at it: presently the field will disappear underneath your nose; and unless you continually steepen your dive you will overshoot that field by a mile!

To make a truly no-lift dive, a truly vertical descent, an airplane must be put quite distinctly over on its back, past vertical. In this condition the wing meets the air at the no-lift Angle of Attack. The wings then create no lift, but only drag. This is the flight condition in which any airplane will achieve its greatest speed.

One should not even talk about dives, however, without pointing out that civilian airplanes and even some military airplanes are not built to stand an excessively fast dive. Quite apart from the severe stresses of the pull-out from the dive, the dive itself imposes great stresses on many parts of the airplane—and some such part might let go. Or "flutter" might develop on wings or tail surfaces. Or the airplane might hit some rough air and break up. Hence every airplane has a red mark on the air-speed indicator, or a placard on the instrument board, which says "Never exceed ——— m.p.h." This placard means what it says, and it means *you*.

THE NORMAL GLIDE

A more useful gait is the Normal Glide. This is shown here with all angles greatly exaggerated to bring out more clearly the whole configuration of attitude, flight path, Relative Wind, and Angle of Attack. This is the normal manner in which an airplane makes its descent to a landing—at least a small airplane.

Of the various power-off descending gaits, the normal glide is the one in which an airplane will cover the greatest distance, horizontally, from a given altitude. If your engine should quit and the best available landing field be rather far away, this is the gait in which you would try to "stretch" your glide to that field.

An entire chapter will be devoted to the management of the airplane in the glide. Just here it is the flight condition itself that interests us. The airplane is nosed down and is kept going by the pull of gravity. Since it is not nosed down steeply, it goes rather slowly; since it goes slowly it flies at a quite noticeable Angle of Attack. Note how the Relative Wind blows *upward*-backward at the airplane, rather than simply backward as it does in level flight. This is of course because the airplane is moving *downward*-forward in a glide, instead of simply forward as it does in level flight. The forward-downward motion of the airplane through the air produces an upward-backward impact of the air upon the airplane.

Note how the airplane's nose points but slightly down, while its actual flight *path* goes down much more steeply. Remember that the Angle of Attack can be defined as "the difference between where the

A normal glide.

airplane *points* (in the up-and-down sense) and where the airplane *goes*." Imagine how this will feel to the pilot: his nose is pointing at a distant field; but as he watches his actual progress he notices that he is coming down far short of the field. This gives him a sensation of "sink." This "sink" is nothing unhealthy or irregular or dangerous; it is nothing but the Angle of Attack, become visible!

Note, finally, that the "elevators," the flippers on the airplane's tail, are deflected upward and that the pilot is holding the stick back. It is often thoughtlessly said that the flippers are the airplane's up-and-down control; that the pilot makes the airplane climb by pulling the stick back, and makes it descend by pushing the stick forward.

A special chapter in this book will point out that the elevator is actually the airplane's Angle of Attack control, and that its up-and-down control is the throttle. The airplane in a normal glide is going down neither "*because*" the pilot is holding the stick back, nor "*although*" he is holding the stick back. It is going down because the throttle is closed! The position of the stick, the upward deflection of the flippers merely fixes the Angle of Attack and the air speed at which the airplane flies as it descends. Because the stick is held back—and the flippers are deflected upward—the airplane flies rather slowly and at rather large Angle of Attack.

THE MUSHING GLIDE

Another gliding gait is the Mushing Glide. It is simply descending flight at very low air speed and very high Angle of Attack. This gait is sometimes used by a skilled pilot, during an approach to a landing, to steepen the descent and at the same time make sure that the airplane will not pick up excess speed, as it would if the descent were steepened by diving. An unskilled pilot sometimes gets into this flight condition inadvertently, and with exactly the opposite intention: while in a Normal Glide he tries to "stretch" his glide (make it more shallow so it will reach farther) by pointing his airplane's nose less steeply down than the Normal Glide requires. He then gets exactly what he does *not* want, for as the airplane slows up and goes into a Mushing Glide the larger Angle of Attack means that its descent actually steepens! This flight condition is also much used by soaring pilots. A sailplane—in fact any normally shaped airplane in a glide—will lose the fewest feet of altitude per *minute* when flown at this gait. (It loses the fewest feet of altitude per *mile* when flown at a Normal Glide.)

Note that the elevator is sharply up, and the stick held well back: that's what causes the slow speed and high Angle of Attack at which the airplane is proceeding. With the airplane's nose pointing at the horizon while its actual flight path is so steeply down, the pilot has, of course, a very lively impression of sink or "mushing."

Sooner or later most pilots discover that a clean (that is, well-streamlined) airplane can maintain a Mushing Glide with its nose pointing slightly *above* the horizon. This is contrary to "reasonable" expectation; one would expect the airplane to get no forward pull at all from gravity when it is held nose-high. One would expect it to lose all speed, and stall. And because it seems contrary to first-glance common sense, pilots frequently argue themselves out of their own observation, and come to think that it is just an optical illusion. It is not. There isn't room here to explain *why* a clean airplane can glide slightly nose-high, but it seems worth stating *that* it can do so.

A slow mushing glide.

THE POWER-OFF STALL: A STUMBLE

If the various glides and climbs are gaits, the stall must be called a stumble. It is not a flight condition, but is the contrary of flight: the airplane stops flying and starts to fall.

Our picture catches this stumble at the exact moment when it occurs. It is a snapshot, as it were, taken at just the critical moment. The other pictures in this chapter illustrate steady conditions which the airplane can maintain indefinitely: this picture shows a condition which lasts only a moment.

Just *before* this picture was "taken" the airplane was in a very slow, very "mushing" glide. The pilot then attempted to fly the ship too slowly, at too high an Angle of Attack, and this brought on the condition shown (with the Angle of Attack greatly exaggerated) in our picture.

A power-off stall.

Note that the airplane *points* slightly above the horizon, but that its actual path is steeply down. Remember that the Angle of Attack is "the difference between where the airplane *points* and where (in the up-and-down sense) it *goes*." The Angle of Attack in this case is excessive. This is not the *result* of the stall, but its *cause!* Meeting the air at this excessive Angle of Attack, the wings no longer succeed in deflecting it downward; they merely disturb it. Hence they make no lift, and the airplane begins to fall.

Note how the pilot got into this trouble: he holds the stick hard back, and the flippers on the airplane's tail are sharply up. It is this upward deflection of the flippers that forces the ship to assume the excessive Angle of Attack. And it is the excessive Angle of Attack that causes the burbling of the air flow on the wings, and the collapse of lift.

Immediately *after* the moment caught in our illustration, the airplane will drop from lack of lift, as indicated by the flight direction arrow. While it drops, it will also nose down—despite anything the pilot may do to keep it from nosing down. It will thus *attempt* to recover speed, *attempt* to reduce its Angle of Attack. If the pilot then relaxes his back pressure against his stick and thus allows the flippers to return to neutral, the airplane will succeed in making a recovery. It will come out into a steep gliding descent from which it can then easily be brought back to a Normal Glide or a Mushing Glide or any other desired flight condition. If the pilot keeps holding the stick back as the airplane drops and its nose dips earthward, the airplane will *not* succeed in making a recovery. If it is an exceptionally well behaved airplane it will simply nose up into a new stall and will drop again. If it is less well behaved it will go into a spin. In any case, as long as the pilot keeps holding the stick back, he keeps the airplane at stalling Angle of Attack—and hence in trouble.

Unfortunately, as the airplane drops out from under the pilot and as its nose dips earthward, the pilot's "instinctive" reaction will be to haul back all the harder on the stick. If his imagination works with faulty images, if he imagines that the stick is the airplane's up-and-down control, he can hardly help hauling back on the stick. This instinctive reaction will be especially impulsive and uncontrollable if the pilot has failed to sense the coming of the stall, and the stall takes him by surprise.

And *that* is the real danger of stalling: this faulty reaction to the stall, rather than the stall itself. It is quite rare that a pilot is killed simply because he stalled. But it happens with tragic monotony that a pilot is killed because, stalled when he did not expect it, he either fails to recognize the stall for what it is, or fails to control that impulsive desire to haul back on the stick: he clamps the stick back against his stomach in a terrified cramplike effort to hold the airplane up, and thereby makes the stall worse or converts it into a spin.

The airplane, once stalled, *must* go down: only by sacrificing altitude can it regain speed quickly enough—as will be explained in more detail in another chapter of this book. And the airplane's stick *must* be allowed to come forward: the whole trouble started only because the stick was too far back! As long as the stick is too far back, the wings can't make lift.

When an airplane is stalled, nothing—absolutely nothing—will help except this one thing: get the stick forward. You might think, for example, that you could regain speed by slamming your throttle wide open; but it will not help as long as the stick is back. It would merely convert the stall into a slightly more vicious power stall, or the spin into a power spin: you would keep going down, out of control, just the same. It is different once the stick has been allowed to come forward and the airplane's Angle of Attack is reduced: *then* wide-open throttle makes it possible to complete one's recovery with quite small loss of altitude. But the stick must come forward *first*.

In stall-and-recovery practice student pilots are sometimes required to keep the stick hard back for a while, delay the recovery and meanwhile practice how to maintain a semblance of control over a stalled airplane even while it drops, so as to keep it from spinning. It consists of a continuous, quick, juggling sort of footwork on the rudder. But even if it isn't spinning, the airplane is still stalled and still dropping and is bound to get away from the pilot and to spin as soon as he makes his first mistake in his juggling, which won't be long. Thus it is not really the answer to the stall. When you are stalled, there is one thing that can really help you: accept the inevitable loss of altitude and get the stick forward.

It has been stated above that the stall is not a steady gait, but a "stumble"—that an airplane cannot be in this condition except for a moment, and will either recover or go into a spin. This may seem contrary to the experience of some pilot who may claim that he often put his airplane into a gentle stall and then descended in the stalled condition for quite a while before making recovery.

The answer is contained in our picture. Most modern airplanes are so designed that the whole wing will not stall all at once. When the airplane's Angle of Attack becomes excessive, the inner part of each wing (the part near the fuselage) stalls while the outer part (the tip part) still keeps making lift. When the eddies flowing off the stalled portion of the wing strike the airplane's tail, they often shake the whole airplane: and in other ways, too, the pilot can sense that the stalling has begun. But the nonstalled part of the wing continues to give him lift, stability, and control and thus enables him to maintain steady descent. It is really an extreme form of Mushing Glide, rather than a

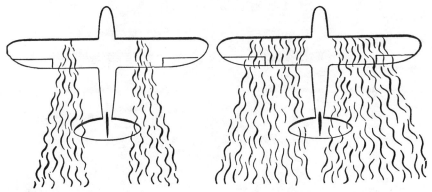

A well-behaved airplane does not stall all at once. As pilot gradually brings the stick back, increasing Angle of Attack, burbling of air flow begins at some certain spot of each wing, usually near the wing root. *Left:* First stage of the stall. If the pilot now does not bring the stick back any farther, the airplane can stay in this condition, maintaining steady flight. An airplane may reach this flight condition in an extremely slow glide. *Right:* If the pilot continues to bring the stick back farther, he stalls additional parts of the wing. As the stall spreads out toward the wing tips, the ailerons become less effective, and the ship more inclined to spin. The airplane cannot be held steadily in this flight condition: it will "drop out" and nose down; but unless the pilot then relaxes the stick, a new and more severe stall will follow.

stall. Part of the wing is stalled. The airplane as a whole is not stalled, though very near the stall.

If an airplane has the opposite stalling characteristics and its wings stall at the tips first, it is tricky to fly. It will stall with little advance warning. The stall will render its ailerons worse than useless and will almost instantly develop into a spin.

THE CLIMB

Our two climb pictures explain the climb better than many words could. Note that the flight path is upward. The Relative Wind blows downward at the airplane! In order to meet this Relative Wind at an

Angle of Attack, the airplane must therefore be flown quite nose-high. If the climb is fairly shallow, the airplane can maintain, at wide-open throttle, a rather brisk air speed and can therefore fly at moderate Angle of Attack. If the climb is steep, the speed will necessarily be slow, since only limited power is available to pull the airplane up that steep slope. And because the speed is so slow only a rather large Angle

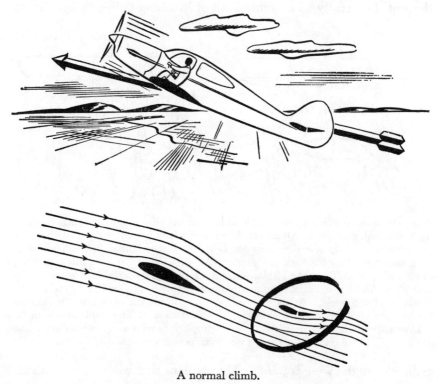

A normal climb.

of Attack can produce enough lift to hold the airplane's weight. Thus the airplane's nose must be carried very high in a steep climb.

The Normal Climb is the gait at which the airplane will gain the most feet of altitude per *minute*. The Steep Climb is the gait at which it will get the most feet of altitude *per mile* of horizontal distance. Since we are usually more interested in climb per minute than in climb per mile, the steep climb is seldom useful. In addition, the steep climb has two disadvantages: being at such high Angle of Attack, the airplane is necessarily near the stall; and because the air flow is slack while the

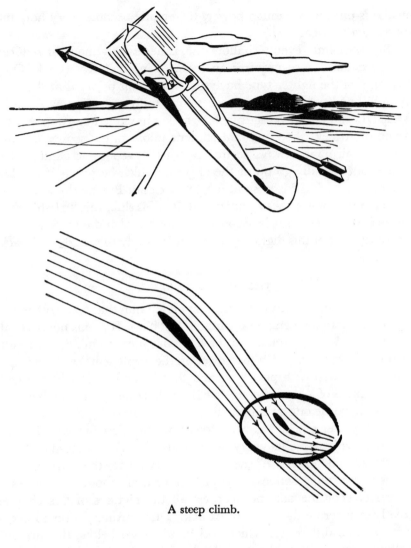

A steep climb.

engine is putting out much power, the engine becomes very hot, and suffers undue wear.

But *any* climb, even a shallow one, is necessarily made at reduced air speed and hence requires an increase in Angle of Attack. Only airplanes of the fighter type have such abundant power that they can fly uphill and yet maintain speed enough to fly at really low Angle of Attack. Because the average airplane makes its climb at an appreciable Angle of Attack, the observant student pilot notices sooner or later that this airplane always seems to "sink" even as it climbs; that it does not actually go up at the angle at which he points it up. For instance, he points his nose at a certain cloud. But he doesn't reach that cloud; he passes 'way underneath it. This sink, this "mushing" of the airplane as it climbs, is of course not abnormal or dangerous—it is a necessary part of this flight condition: it is simply the Angle of Attack, visible to the eye.

THE POWER-ON STALL

The layman is always surprised if you tell him that an airplane can stall even though its engine is running full blast. It seems nonsensical, yet it is so. With remnants of the layman's ideas in mind, many student pilots think that with the power on the airplane has at least much more "lift" than with power off and that it can be brought to much higher Angle of Attack before it will stall. It seems reasonable, and *seems* to be borne out by experience; yet it is not so.

With power on, just as with power off, the airplane stalls for one reason only: because its wings are meeting the air at excessive Angle of Attack. And "excessive" means the same whether the power is on or off: for most wing shapes, anything more than about 18 degrees is excessive. (For the sake of clearness all the pictures of this chapter greatly exaggerate the Angle of Attack.) Disregarding for the moment certain small differences which will be discussed below, the airplane stalls with power on at the same Angle of Attack at which it stalls with power off.

Our picture catches the moment when a power stall occurs. Tilt the power stall picture over on its side so that the flight direction arrow points downward and the airplane's nose points only slightly up: you will see that the whole configuration of flight path, attitude, and Angle of Attack is then exactly the same as in the power-off stall picture. The

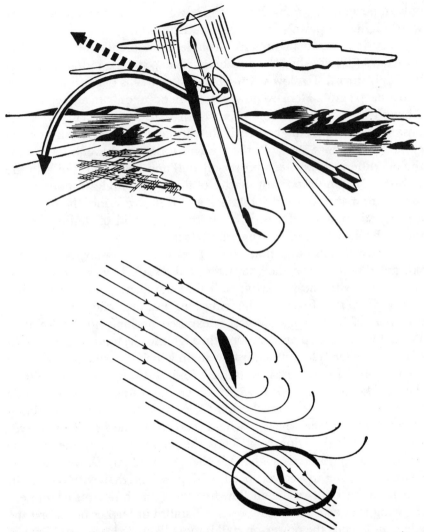

A power stall. Because all these pictures (illustrating the gaits of the airplane) are exaggerating the Angle of Attack—and because the pictures try to be consistent with one another, this picture is rather extremely exaggerated: Angle of Attack is much greater than it could ever be in reality, where 18 degrees is maximum; hence airplane's attitude is more extremely nose-up than it could be in reality in a gradually approached power stall.

only important difference is that the power-off stall happens in down·· ward flight, the power-on stall in upward flight.

Why in upward flight? An airplane cannot stall unless it is first slowed up (disregarding the possibility of snap stalling it or of stalling it in a tight turn). To slow it up when the power is on, it must be put into a steep climb. Hence a power stall can occur only in a steep climb! That's why the nose is so extremely high in such a maneuver: the nose would be very high even if the airplane were simply in a steep climb. To produce the stall, the nose must be raised even more. Thus the big difference between power-off stall and power-on stall is in the comparatively unimportant matter of flight path and attitude; in the important matter of Angle of Attack and Relative Wind, the two stalls are essentially alike. There is really only one kind of stall: the wing meets the air at excessive Angle of Attack.

It must be admitted that power does make certain small but important differences in the character of the stall. It was explained in connection with the power-off stall that the average airplane's wing will begin to stall first near the wing root, last near the wing tip. The root part of the wing, however, is exposed not only to the Relative Wind due to the airplane's motion through the air, but also to the blast of the propeller. The actual air flow upon which the wing root works is a composite of two winds, Relative Wind and propeller blast. When the airplane as a whole is brought to very high Angle of Attack while the power is on, the propeller blast keeps the air flow from burbling at exactly the place where it would tend to burble first if the power were off: near the wing root. Thus the airplane as a whole may be brought to slightly (*very* slightly) higher Angle of Attack (and thus be flown at very slightly slower speed) with power on than with power off. When the burbling of the air flow finally begins, it doesn't begin near the wing root as it does in a power-off stall, but farther out along the wing, and hence the power-on stall is more likely to lead to sudden loss of lateral stability and loss of control: the airplane is likely to capsize sideways as it stalls; and if the pilot then misuses his ailerons, the stall is more likely to develop into a spin.

This stall-delaying effect of the propeller blast is particularly marked in multiengined airplanes, because in such airplanes larger parts of the wing are exposed to propeller blast. Thus such airplanes can be flown quite appreciably more slowly, and at appreciably higher over-all Angle of Attack with power on than they can be flown with power off. But even with such airplanes, these differences between power-off and power-on stall are small, as compared with the similarities. Essentially, there is only one kind of stall: the wing meets the air at too much of an Angle of Attack. There is only one way to get an airplane into a stall: the stick is pulled back, the flippers are deflected upward and thus force the airplane to high Angle of Attack. And there is only one way to get an airplane out of a stall: get the stick forward.

THE POWER DESCENT

The nose points *up*, but the airplane goes *down*. This flight condition is not used much in primary flight training; but it is the best way to make a landing approach in a heavy, powerful airplane.

A power descent.

Our picture shows what goes on. It is simply a slow mushing glide in which just enough engine power is being used to make the airplane not go down so steeply. The Angle of Attack is the same as in a glide, but the flight path does not go so steeply down; hence, in order to keep the Angle of Attack high and the speed slow, the nose must be carried quite high.

How and why this peculiar gait is used will be discussed in the chapter on Landings. How the airplane is controlled when making a power descent will be explained in the chapter on The Flippers and The Throttle.

FLIGHT UNDER *g* LOAD

Whenever an airplane's flight path is curving, the airplane becomes heavier than it was in straight flight: it loads itself down, as it were, with centrifugal force. At least that is one way of describing the phenomenon which pilots call *g load*. Because of this additional load which crowds itself on, the airplane's flight condition in curving flight is different from its straight-flight condition and must be considered as yet another distinct gait of the airplane.

This *g* load, this pull of centrifugal force, is of course not peculiar to airplanes alone. It acts upon all things which are in curving motion. It merely is particularly noticeable in an airplane because the airplane moves so fast and because the pilot is so extraordinarily free to make his flight path curve crazily right or left, up or down as he chooses. Hold a 1-pound chunk of wood suspended from your hand by a string. It pulls on your hand, of course, with the force of 1 pound. Whirl that same weight around, so that it describes a circular path, and it pulls on your hand with a force of many pounds. This additional pull is not really additional weight but is simply centrifugal force, but it pulls on your hand exactly as if it were real weight.

When an airplane goes around a curve, the same thing happens. The only difference is that the pilot himself is now part of the whirling weight, instead of being at the stationary center, and that he therefore *feels* the effect in a different fashion: he himself becomes heavier too! This increase in one's weight is quite confusing to passengers and beginning students! But the thing to remember is this: just as the pilot becomes heavier so does, of course, the whole airplane become heavier in a turn: in a 45-degree banked turn, for example, centrifugal force has the same effect as if 40 per cent had been added to the airplane's weight. In a 60-degree banked turn, centrifugal force has the same effect as if the airplane's weight had been doubled: in pilots' language, two *g*'s are then acting on the airplane.

In curving flight, then, the airplane's wings must support not only the ordinary weight of the airplane but also this *g* load. Hence when an airplane, flying at a given speed, describes a curve it must fly at an *additional* Angle of Attack in order to create the *additional* lift necessary to support this *additional* "weight." The pilot, while flying a curve, produces this increased Angle of Attack by putting back pressure on the stick, thus deflecting the flippers upward. Should he fail to do so,

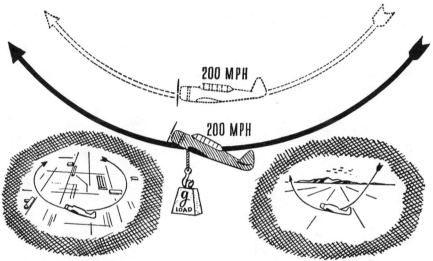

Flight under *g* load: *Inserts* show maneuvers which produce *g* load: turn, and pull-out from dive. *Dotted sketch* shows how this particular airplane *would* fly at this particular speed, if it were not for *g* load. *Solid sketch* shows how it *does* fly because of *g* load. Note upward-deflected flippers: pilot holds stick back to maintain large Angle of Attack.

the curve would simply not materialize, or the airplane would go into a dive, while curving, under the "weight" of the *g* load.

Consider what this means. In *any* turn, the airplane "mushes"; that is, it flies at high Angle of Attack. Even though it may have a speed at which in *straight* flight it could fly at low Angle of Attack, it will mush in the curve. If the curving of the flight path is at all sharp, the airplane may under the squashing "weight" of the *g* load, even reach stalling Angle of Attack—even though flying at quite considerable speed! When flying a very steeply banked, very tight turn, for example, an airplane whose straight-flight stalling speed is 50 m.p.h. will stall at 80 m.p.h.! In short: an airplane in curving flight is a heavily overloaded airplane and behaves as such.

It makes no difference in this respect whether the curve is a curve to the right, to the left, or upward! The pull-out from a dive is in this respect precisely the same thing as a sharp turn to one side or the the other: centrifugal force adds itself to the airplane's weight and overloads the airplane. The airplane will "mush" in such a pull-out and may even reach stalling Angles of Attack, even though its speed may be very high, and even though its nose may be pointed steeply down.

Chapter 3

LIFT AND BUOYANCY

LIFT is one of the most confusing things in flying to talk about. On no other subject is there so much difference between what the engineer claims and what the pilot knows. Every pilot knows, for instance, that an airplane has more lift in fast flight than in slow flight. And when an airplane mushes, he knows it has very little lift indeed; that's why it mushes! But the engineer calls that plain ignorance. He claims that an airplane's lift is always the same (except for momentary disturbances) regardless of whether it is flying fast or slowly, and practically regardless of whether it is flying uphill, downhill, or level. The pilot answers that this may be so "in theory," but then theory is only theory, that is, bunk. In practice, you just try and come in for your landing a little too fast, and you'll find out—why, your excess lift will float you clear across the airport! At this point, the engineer starts talking about Angle of Attack again, and the pilot's attention begins to wander.

The trouble is purely vocabulary; pilots and engineers don't speak the same language. The word *lift* means one thing to the engineer and another thing altogether to the pilot. Each of the two meanings, however, describes something real; and since lift, whatever it may be, is essential for flying, it is worth while to figure out just what each means by lift and just what each knows about it.

THE ENGINEER'S IDEA OF LIFT

The engineer has a clear-cut definition of lift. Lift is that component of the total air force acting on a wing which acts at right angles to the direction of flight. In straight and level flight, this means simply the upward force the air exerts on the wing. In normal glides and climbs, it means practically the same thing. The engineer makes one clear-cut statement concerning lift: in steady flight, the lift of an airplane is always equal to its weight—regardless of the speed of flight, and practically regardless of whether the airplane is flying level,

45

climbing, or descending.* If a 2,500-pound airplane is in steady flight, that is, neither ballooning nor dropping nor looping the loop nor flying a curve, then its wings are developing 2,500 pounds of lift; not 2,499, not 2,501, but exactly 2,500.

This idea, properly understood, throws much light on the art of flying. There are two ways of understanding it. Perhaps one should read up on elementary physics—on the laws of motion, on gravitation, on the behavior of falling bodies, on the difference between steady motion and acceleration. Certainly a flight instructor should understand all that to keep from talking nonsense. It may be just classroom sort of theory for people who work at desks, but it is practical insight into everyday affairs for people who work in a cockpit, moving in three dimensions and with six degrees of freedom.

AN EXPLODED PICTURE

But one can understand this thing—the equilibrium between an airplane's lift and its weight—also in a more direct way by observing closely how an airplane behaves in flight. You have seen "exploded" drawings of a machine, pictures in which the machine is slightly pulled apart so that you can see its individual parts and yet can see also how they belong together. Well, one can "explode" a complicated thing in time as well as in space. Here is an "exploded" story of how an airplane in flight always adjusts its own lift to its own weight.

Start out with the airplane flying straight and level at cruising speed. Obviously, its lift and its weight are then in equilibrium. If they were not, the airplane would not fly steadily. If it had more lift than weight, it would balloon upward. If it had more weight than lift, it would start sinking. The very fact that it does neither shows that it is in balance.

* "Practically." The proposition is oversimplified here. In strictly correct form, it reads: In steady flight "the sum of all upward forces" (rather than "the lift") equals "the sum of all downward forces" (rather than "the weight"). This takes account of the fact that, in a climb, with the airplane pointed slightly up, the pull of the propeller is directed slightly upward and thus helps slightly in holding the airplane up as well as pulling it forward. Also, the drag is directed slightly downward and tends to hold the airplane down a little, as well as holding it back; and the lift is no longer straight up but acts slightly backward. Similar considerations apply to the glide. But as long as we consider only level flight, normal climbs, and normal glides, we may disregard these refinements and assume that the only important upward force acting on the airplane is its lift and that the only important downward force acting on it is its weight. It makes the discussion much simpler and still carries the main idea.

Now, let the pilot reduce speed by, say 15 m.p.h. At the same time let him keep the airplane's attitude strictly as it was before—doing with the controls whatever is necessary to keep the nose, relative to the horizon, exactly where it was before. And now remember that this is an "exploded" account, that we must pull apart and present *as a long-drawn-out time sequence* a number of events that really happen almost simultaneously as causes and instantaneous effects. Assume, then, that something happens that in reality has not time to happen—that for a few moments the airplane's flight path remains level. What would be the flight condition of the airplane during those few moments?

The flight path, during those moments, is the same as before. The airplane's attitude is the same as before. Hence the relative wind blows at the wings from the same direction as before. Hence, the Angle of Attack of the wings is the same as before. Only the speed has changed. And because the wings now move through the air at a slower speed, they now develop less lift. The 2,500-pound airplane thus is supported, during those few moments, by only perhaps 2,000 pounds of lift force.

What happens next?

The airplane behaves exactly as any object will behave which weighs 500 pounds and finds itself unsupported in mid-air—it drops. It does not nose down, because in this flight experiment the pilot does not allow it to nose down. If it nosed down, it would pick up new speed, and the whole experiment would be over. No, the airplane simply begins to fall through, flat. It does not merely "descend"; it *falls*, and there is a difference. The elevator of a building descends, that is, goes down at a nice even speed. But a stone dropped down the elevator shaft falls; it goes down slowly first and then faster and faster, because gravity keeps pulling on it and accelerating it downward, and gravity is not opposed by any other force. The airplane that has not quite enough lift for its weight behaves like a falling stone; it goes down slowly at first, and then faster and faster, picking up *additional* downward speed every moment because gravity keeps tugging at those 500 pounds of unsupported stuff. If nothing else happened to stop the airplane's falling, it would finally crash into the ground with a terrific thud—simply because the pilot reduced speed by 15 m.p.h.!

Actually, something does happen to stop its falling—happens in fact so promptly that it falls only a few feet, or a few inches. As the airplane begins to sink, it finds a cushion of new lift; and what creates this cushion is the airplane's sinking! Sinking creates lift.

SINKING INCREASES THE ANGLE OF ATTACK

Here is how this happens. As the sinking begins, the Relative Wind begins to change. Since the airplane now moves downward through the air as well as forward, the Relative Wind now blows upward at the wings as well as from in front. This means that the Angle of Attack becomes greater. Even though the airplane's attitude still does not change (the nose is still pointing strictly level at the horizon), its Angle of Attack does change. Remember, Angle of Attack is the angle at which the wing meets the *air*. The greater Angle of Attack means that more lift is made. The loss of speed has destroyed some lift; now the increase in Angle of Attack restores that lift. The more sharply the airplane sinks—with its nose held level— the greater becomes the Angle of Attack, and the more additional lift is created. Presently the airplane reaches a sinking speed at which the Angle of Attack is so high that the lift is again equal to its weight; and from that moment on it no longer behaves like a falling stone, but again like an aircraft; it is again fully supported by lift.

Just where will the airplane go from then on? A quick first guess would be that it will fly level. With lift again equal to weight, it seems that it should. More careful thought shows that the airplane will simply stay in the flight condition in which it has found equilibrium; it found the new equilibrium in descent, and in descent it will stay. But it will be a steady descent, not the falling, accelerating sort of descent. The ship will be in the flight condition described earlier as a "power descent."

The same process also works the other way 'round. Whenever for any reason the airplane has more lift than it has weight, it balloons upward; but by ballooning upward, it changes its own relative wind and thus reduces its own Angle of Attack; the lift decreases and becomes again equal to the weight. The airplane thus regains equilibrium in climbing flight; and, until the pilot does something about it, it will stay in climbing flight.

This happens, for example, when a bomber drops its bombs. With so much weight gone there is then suddenly a lot of excess lift, and the airplane surges upward. The pilot can kill the excess lift by getting the stick forward, thus reducing the ship's Angle of Attack, or he can kill it by throttling back; but if he does nothing but simply hold the ship in the same level attitude, here is what happens. The excess lift causes the ship to balloon upward; and the ballooning then reduces the ship's Angle of Attack because it makes the relative wind blow at it from slightly above. Thus the lift is reduced, and the ship steadies in climbing flight.

ROUGH AIR

Much the same sequence of events happens when an airplane flies through rough air—except that then the original disturbance of weight-lift equilibrium is not caused by a change of air speed or by a change of the ship's weight but by a change in Angle of Attack. This is worth understanding.

Suppose an airplane flies into an updraft. When the upflowing air first meets the wings, the Angle of Attack becomes temporarily rather high because the air is then flowing upward against the wings instead of flowing at them level from in front. Owing to the increased Angle of Attack, a tremendous surge of lift develops. The airplane balloons upward; the occupants feel the ship come up against them with an upward jolt that makes them feel heavy in their seats; but this upward motion of the airplane makes the wind of flight blow at the wing from slightly above, thus reduces the Angle of Attack, and thus reduces the lift. Equilibrium is restored, and the airplane continues in steady upward flight as long as it remains in rising air.

Next, suppose it flies out of the region of rising air and enters a downdraft. When its wings first meet the downflowing air, the resulting Angle of Attack is temporarily quite low, the lift is not enough to hold the airplane up, and the airplane sinks: but by sinking it increases its own Angle of Attack and thus restores its own lift; it steadies down in descending flight, and it keeps descending steadily until it has passed through the downdraft—or until the pilot does something about it. And all the time, both in the updraft and the downdraft, the attitude of the ship was strictly level.

Thus an airplane always seeks that flight path which will result in equilibrium between its weight and its lift. Remember again, these have been "exploded" accounts. Actually, the airplane does not jump around roughly except in violent upgusts and downgusts. Normally, the adjustment of lift to weight is continuous and smooth. The equilibrium can rarely be off by hundreds of pounds, because the whole restoring process goes to work as soon as it is off by only a few ounces.

WHEN THE BOTTOM DROPS OUT

All this throws additional light also on an airplane's behavior in the stall. When the wings are stalled, the equilibrium-seeking process just described will not work.

Consider: when a wing is stalled, an increase in Angle of Attack will lead not to more lift, but to less lift. That is in fact the proper definition of a stall—"a flight condition in which further increase in Angle of Attack leads to a decrease in lift." Hence, when a stalled airplane begins to sink, it does not thereby restore its own lift, as an unstalled airplane would; the increase in Angle of Attack that results from the sink destroys its lift! This causes an even faster sink; the faster sink, by increasing the Angle of Attack still more, causes an even more severe reduction of lift; and so on, in a vicious circle. Where the unstalled airplane would, by sinking, find a cushion of new lift, the stalled airplane falls right on through, faster and faster. That's what causes that well-known feeling, in a stall, that the bottom is dropping out from under you: the bottom *is* dropping out from under you!

WHY AN AIRPLANE SPINS

But that is not all. What's true of the airplane as a whole is true also individually of its right wing and its left wing. In healthy flight, when one wing of the airplane dips down, it thereby always finds a cushion of new lift, which then tends to stop it from going down farther; and at the same time the other wing, by going up, reduces its own lift and thereby keeps itself from going up farther. This so-called "lateral damping effect" is an important ever-present factor that helps a lot to make our airplanes well behaved and steady laterally when at first glance one might expect them to be cranky, like canoes.

A pilot should clearly understand this effect. What causes that steadying cushion of extra lift is not the fact that the wing *is* down, but that it is *going* down. The downward motion causes a Relative Wind which blows upward; out at the wing tip, where the downward motion is quite brisk, the upward Relative Wind is also quite brisk. A student can understand this best on the ground, by stretching out his arms as if they were wings, and then rapidly "banking" and "unbanking" them: he will feel the resulting Relative Wind on his palms and on the backs of his hands. When such an upwardly directed Relative Wind, due to the wing's going down, combines with the backwardly directed Relative Wind due to the airplane's forward motion, then there results a *total* Relative Wind which blows slightly upward against the downgoing wing. Thus whenever a wing dips down, it always thereby increases its own Angle of Attack. The extra Angle of Attack thus causes an extra lift; the extra lift tends to stop the wing from dipping down farther. As soon as the wing has stopped going down, however, the whole effect disappears; it does not tend to bring the wing back up. Thus it does not *stabilize* the airplane, does not tend to restore it to normal condition; that is done by other effects, to be discussed later in this book. The effect discussed here merely "damps" the airplane's motions, makes it steadier, less cranky and sensitive, less ready to roll, and thus much easier to control.

But, once the airplane is stalled, this effect reverses itself! When one wing goes down and thus increases its Angle of Attack, it thereby finds *no* cushion of new lift, but on the contrary it destroys even more of its own lift and keeps wanting to go down. At the same time the other wing, in going up, reduces its Angle of Attack and may thereby actually unstall itself, gain lift, and keep wanting to go up! Thus a stalled airplane is laterally unstable and undamped; the more a wing drops, the more it wants to drop.

That is how a stall may become a spin. Both wings are stalled. One wing dips down; by going down it stalls itself still more and thus keeps going down and thus keeps itself stalled. At the same time the other wing keeps having more lift, going up, and getting even more lift. Engineers call this process *autorotation*. Let it happen on an airplane in flight; let the centrifugal force go to work; add the effects of different amounts of drag on the two wings—and you get the kind of motion known as a spin.

THE PILOT'S "LIFT": BUOYANCY

But now, what does the *pilot* mean by "lift"? The pilot does not mean "lift" at all, in the engineering sense of the word. He means a certain somewhat mysterious something that you can sense when you sit in an airplane and fly it. Sometimes the airplane has it; at other times it hasn't. Whether or not it has it is of the greatest importance to the pilot.

What is this mysterious quality? Unfortunately it has no name, (other than being misnamed "lift"). Most things become less mysterious if you can give them a name. One might call it the "flyingness" of the airplane—how far at any moment it is removed from a stall; or, if you cut its throttle, how far it could keep floating without loss of altitude. That is what the pilot is interested in when he comes in for a three-point landing. "Have I got too much lift?" means, "Am I going to float too far across the field before she is willing to stop flying and to sit down and to *stay* down?" That's how a beginner first becomes conscious of this whole matter of "lift." Coming into the airport with a little too much "lift," he finds that the airplane is unwilling to stay on the ground. Even if its wheels are put in contact with the ground, it simply bounces off the ground again and keeps flying. It may float clear across the airport, sitting down just in time to run into the fence!

FIRMNESS OF SUSTENTATION

Or one might call it "Firmness of Sustentation." That's what the pilot is interested in when he climbs steeply over some obstructions or when he has slightly misjudged his glide and is trying to "stretch" it. "Have I got enough lift" means, then, "Is she going to stop flying and settle out from under me before I have crossed the airport boundary fence?" This, too, is a problem for all pilots; for in an attempt to avoid an excess of this so-called "lift," one may easily try to "come in" with too little. And while the landing approach with too much "lift" may mean damage to one's ship and reputation, an approach that suddenly runs out of "lift" means that the wings suddenly let go in a stall, and may then mean a nose-first dip into the ground, and an end to all further problems. That's why it is so important for the pilot to have just the right amount of this so-called "Lift"—neither too much, nor too little.

EXCESS SPEED!

One might of course call this thing simply: "Excess Speed over Stalling Speed," and be done with it. "I have lots of lift" would then mean: "I am flying much faster than the minimum speed absolutely necessary to keep this ship in the air; I can afford to cut all sorts of capers, lose a lot of speed before I shall have to worry about stalling." "I have very little lift" would mean: "I am flying so slowly that the least further slowing up would cause me to stall." This is a good explanation to use in the hurry and noise of actual flight instruction. But it is not quite enough. The explanation is simple and convincing merely because it does not really explain. Pilots say: "In slow flight an airplane has very little lift"; and they really *mean* something by that statement. Translating "lift" as "Excess Speed over Stalling Speed," the statement would become: "When flying slowly, you are flying at a speed which is only very little faster than stalling speed"—which is saying almost nothing at all, and still leaves open the question just what this pilot's sort of "lift" really is. In addition, remember an airplane has no *one* stalling speed. The speed at which an airplane will stall varies with its load, with the maneuver in which it is engaged, and with air conditions. Thus 60 m.p.h. may be excess speed, and then again it may not be!

THE ZOOM RESERVE

Continuing, then, our attempt to understand this mysterious something that pilots call "lift," we might try to call it "The Zoom Reserve." A zoom is a very steep climb, so steep that the airplane can't hold it as a steady condition but gradually slows up. As it slows up, the pilot must of course increase its Angle of Attack by bringing the stick back; otherwise the zoom would level out and stop. Thus a zoom, continued long enough, will always end in a stall. "I have lots of lift" may simply mean, then: "I have enough speed, and am flying at low Angle of Attack, so that if I started a zoom at this moment, I could make it long and steep before I would stall." "I have very little lift" may mean: "I am in a flight condition which would lead to stall almost right away if I now started a zoom."

POTENTIAL EXCESS LIFT?

Again, one might call this mysterious something the "Potential Excess Lift." In *steady* flight, an airplane's lift (real, engineering sort of lift, that is) is always just equal to its weight. But by pulling his stick back and thus holding his wings at higher Angle of Attack, the pilot can produce a surge of excess lift. The excess lift will then make the airplane's flight nonsteady: it will balloon the airplane upward in an up-curving, zooming flight path—perhaps the pull-up into a loop. "I have lots of lift" then means: "I am in a condition where, simply by pulling my stick back, I could double or triple my lift force if I wanted to, and could make this airplane shoot upward as if it had been given a tremendous kick from below." "I have very little lift" means: "I am in a flight condition where I could get only a feeble surge of excess lift, and a feeble upward curving of my flight path even if I now pulled my stick all the way back." And "I haven't any lift at all" would mean: "I am in a flight condition where, if I pulled the stick back a little farther, no additional lift would be produced, but on the contrary the wings would stall and the lift would be reduced and the ship would fall."

LOWNESS OF ANGLE OF ATTACK

Thus one could also call this mysterious thing "Remoteness from the Stall." "I have little lift" means: "I am flying near the stall condition." "I have lots of lift" means: "I am far from stalled."

Or one might call this thing, awkwardly but accurately, "Lowness of Angle of Attack." For that's what it really is; that's what the engineer would call it.

What the pilot is concerned about is not really his lift. The very fact that his airplane is not dropping out of the air proves that he *has* lift. What he is concerned about is the Angle of Attack which gives him that lift. When he says he has "lots of lift," he means that he is flying at low Angle of Attack; he is flying fast, and a low Angle of Attack is sufficient to produce enough lift to hold up the airplane's weight. When he claims that he has "very little lift," he means that he is flying at high Angle of Attack; he is flying slowly, and it takes a large Angle of Attack to produce enough lift to hold up the airplane's weight. In normal maneuvering, he wants "lots of lift": he wants to fly at low Angle of Attack because that means the airplane will be efficient, maneuverable, fast, and safe. But when approaching for a three-point landing—which is really a stall brought about when the airplane is flying at 6-inch altitude above the ground—then he wants "not too much lift": he wants to fly at a fairly high Angle of Attack so that a small additional increase in Angle of Attack, brought about by a small additional pull-back on the stick, will get his wings to the stalling angle.

That, then, is what one really ought to call this mysterious thing which pilots like to call "lift"; the only really correct expression for it is "Lowness of Angle of Attack." But the expression of course is too clumsy for practical use.

In this discussion we shall arbitrarily call it *buoyancy*.

The moment you say "buoyancy" instead of "lift," all the pilot's statements turn out to be perfectly true, and the engineer can take no exception to them. It is true that in fast flight—when the wings operate at low Angle of Attack and are far removed from the stall condition—the airplane has much more buoyancy than in slow flight. It is true that in "mushing" flight the airplane has almost no buoyancy. And it is true that, if you come in for a landing a little too fast, your excess buoyancy will carry you a long way across the field. Once you understand that what the pilot usually calls "lift" is really buoyancy or Lowness of Angle of Attack, the whole puzzle simply disappears.

But it is interesting to note that the pilot feels the need of a word for this thing—even though the word may be ill chosen. It shows that Angle of Attack is after all not a "theoretical" concept; it proves that even to the pilot—just as to the engineer—it is the central thing of all flying. You talk to a pilot about Angle of Attack and you get only a blank stare. Yet, a moment later he talks back to you about what he calls "lift," and what we here shall call "buoyancy," and how it makes all the difference in flying: and—since buoyancy is nothing but Lowness of Angle of Attack—what he is really talking about is Angle of Attack!

Chapter 4

THE FLYING INSTINCT

THE important thing to study about buoyancy is not what it consists of, for that is obvious; it consists of Lowness of Angle of Attack. The important thing about it is how it can be sensed.

For this is perhaps the most hard-to-get-at skill in the whole art of flying—the sensing of "lift," the gauging of the firmness of one's sustentation, the "feel" a pilot must have for his ship's Angle of Attack, the ability to know how close the ship is to the stalled condition: this is what pilots used to call the *flying instinct*.

The pilot needs this sensing of buoyancy during almost every minute of almost every maneuver. He needs it especially every time he makes an approach for a three-point landing, for a three-point landing is made with the airplane stalled (or nearly so), and thus the airplane must come quite close to the stall even during the approach. And how many times have you leveled out for a landing only to find yourself with a lot of buoyancy left that you had not suspected—and no way to get rid of it in a hurry? The pilot needs this sense of buoyancy also when climbing out of a tight airport; and, though he sometimes does not realize it, he needs it also during a turn. His life depends on his ability to sense "lift" or the loss of it; most accidents happen only because the pilot's sensing of his buoyancy failed him, and he stalled or spun.

The student pilot realizes all that in a vague way. He feels that he needs this sense; but he also knows that he does not have it. He realizes that he must not stall; but he also realizes that he cannot tell how close he is to the stall. And so he feels like a blind man walking around near a precipice.

The instructor, strangely enough, is rather silent on the subject, and "the book" disregards it almost entirely. The training manuals for pilots pay little attention to it, and the Theory of Flight books none at all. This probably accounts for much of that vague feeling of frustration which bothers most student fliers during training. The

56

student feels that something important is being left unsaid, that only a few questions asked, a few answers given, would help him over the hump; but he is so little oriented that he can't even ask the right questions.

Therefore, we shall now ask one of the key questions in the whole art of flying: How does the pilot know how far he is from the stall, how much buoyancy he has? And we shall take plenty of time to give a full answer.

SPEED

The most important cue is speed. What we call buoyancy or Lowness of Angle of Attack or Firmness of Sustentation is really *almost* the same thing as speed; fast flight means flight at low Angle of Attack, slow flight means flight at high Angle of Attack. Speed and buoyancy are not always quite the same, however. The Angle of Attack at which an airplane must fly depends also on its load. If heavily loaded, it needs more Angle of Attack (at the same speed) than if lightly loaded, and hence it has (at the same speed) less reserve buoyancy; it is a little nearer to the stall. And in this respect the centrifugal force of a turn or of a pull-up from a dive ("*g* load") arts just as if it were real weight. In a turn or a pull-out, the airplane's heaviness increases, and thus its Angle of Attack increases, and thus its reserve of buoyancy decreases, and it comes closer to a stall even though the speed has not diminished. But in straight flight, speed and buoyancy are the same thing; and if a pilot can sense his speed, he can thereby also sense his "lift," that is, buoyancy.

Speed through the air is what counts, of course, not speed over the ground. The pilot therefore cannot judge simply by looking at the ground, as one judges speed in a car by looking at the scenery. The wind-drift effects would deceive him. A tail wind would give him an impression of high air speed; a head wind, an equally deceptive impression of slow air speed. And at the higher altitudes all visual speed impressions become so uncertain anyway that they are useless.

REASON

Air can't be seen: hence air speed can't be seen: therefore the pilot, just like a blind man, must make his other senses do what his eyes can't do. He can judge first of all by logical reasoning. "My throttle is wide open. My nose is pointed up only a little; therefore, I probably have a fairly good air speed." Or, "My throttle is closed. My nose is pointed well down; therefore, I probably have a good but not excessive air speed." But this sort of logic may be treacherous. For example, on a take-off and climb from a high-altitude Western airport, wide-open throttle will not produce the usual amount of thrust, because the air is too thin. A pilot who points his nose up at the usual angle, trusting his wide-open throttle, would stall. Again, there are certain maneuvers (explained elsewhere) in which an airplane can stall with the nose down. Thus, if a pilot attempts to judge by reasoning, he had better be sure that the reasoning is correct and that it takes all factors into account.

Perhaps the most deceptive of these factors is g load. When an airplane, flying at a certain speed, goes into a turn and loads itself down with g load it assumes larger Angle of Attack and thus gets closer to the stall. That has been described earlier in this book. But it isn't the whole story. At the larger Angle of Attack, the wings have more drag, and thus the airplane will slow up, unless the throttle is opened wider. The airplane thus assumes a *still* higher Angle of Attack and gets *still* closer to the stall! Few pilots realize how strong and dangerous this effect is. The average small airplane, fully loaded and with its throttle set at cruising, is actually unable to hold *indefinitely* any turn banked much more than 45 degrees! The effect just described will slow it down gradually, as it circles, so that the pilot's stick comes farther and farther back; until finally, after perhaps twenty turns have been completed, it will stall: stall, mark you, out of level flight with cruising throttle!

That sort of thing makes any speed judgment by throttle position, engine noise, and nose-position highly dubious.

SOUND

A good speed clue is the sound of flight—especially in a glide. The hissing, howling, whispering of the air on the airplane's skin, in its ventilators, on its wires and struts differs at different speeds and becomes generally lower and softer as the speed slackens. The pilot pays attention both to the pitch of those sounds—indicating his speed —and to their changes—indicating whether at the moment his speed is increasing or decreasing.

Gliders are flown largely by their hiss. The older open-cockpit biplanes used to be flown largely by the sound of their bracing wires; it used to be said that if a student tightened up in one of those and slowed it up dangerously in the glide, the wires would hum down a descending melody, "Nearer My God to Thee"—quite an appropriate stall warning.

Even in modern cabin ships, sound helps. Some pilots will open their cabin window when they start down for an approach, so they can listen better to the sounds. In a ship with which he is familiar the pilot often knows some particular sound that comes only at some particular speed, for instance, at a good approach speed, or at a dangerously slow speed. It may be a rattle, or a slapping of the fabric somewhere, or a buzz. Many a ship gives out a sigh or a short howl just before it "lets go" in the stall. These sounds are purely a matter of chance, of course, and vary from ship to ship. But it has been seriously proposed that airplanes be equipped with some musical instrument similiar to an aeolian harp, on which the relative wind would play different notes at different air speeds!

Even the unskilled ear of the beginning student can easily hear and interpret the sounds of flight. The catch is that the beginner does not pay attention to them. In his ground life he judges speed by eye; and in the air he does not think of judging speed by ear any more than he would try to judge it by smell. Moreover, what he needs most is a warning when he is slowing up too much; but as the speed fades, the sounds fade; and as the sounds fade, they also tend to fade from his attention. The greatest danger signal of all is nothing at all, that is, a silence: it is more easily disregarded than a loud noise would be. In ground life, silence tends to mean security; loud noise, danger. In the air, it is the other way 'round.

But that's what learning to fly largely consists of: the pilot need not develop any new senses but must learn to use his old senses for new and different work. Particularly, he must get away from the domination of his eyes and make much heavier use of his ears, his sense of touch, his inner ear with its feel for acceleration and balance. Often in flying the eye gathers one impression, and the other senses gather an entirely different impression. In a gliding turn, for example, the nose of the airplane may be well down and the pilot's eye may thus be convinced that there is plenty of speed. But the speed may actually be quite dangerously low (because of various reasons, explained just above), and the ear will hear that lack of speed. Thus there may be conflict between eye and ear. Lifelong conditioning on the ground would then lead the flier to believe his eyes: but if he had enough air experience, he will believe his ears.

Unfortunately, the sounds of flight are not a good clue in the flight condition out of which most stall-spin accidents develop—flight with power on. The engine noise then drowns out other more revealing sounds. Engine noise itself does change with speed; if the airplane is slowed up by excessive climb, the engine "labors." The ear can easily detect an increase or decrease of 50 r.p.m. But still, it is possible to stall with the engine turning at about cruising r.p.m., that is, with the engine noise continuing briskly. In fact, a false reassurance caused by brisk engine noise, is probably one of the contributing causes of the usual stall and spin accident.

AILERON FEEL

Power on or power off, the feel of the ailerons serves as a clue to air speed. The slower the air speed, the farther to the right and left must the stick be moved to get results—and the slower are the results in coming. In very fast flight, the stick becomes quite stiff, and if you watched it while you rocked your wings, you would hardly notice any actual stick motion. In slow flight, the stick becomes soft, and if you then tried to rock your wings in the same rhythm as before, you would see the stick moving through a wide arc from right to left. Something similar is true of fore and aft movements of the stick—but this will be discussed later. The rudder, too, goes soft as the speed slackens; but both the rudder and the flippers work in the slipstream of the propellers and thus will feel deceptively firm when the power is on, even though the air speed is perhaps quite low. The ailerons alone work in the true Wind of Flight, unaffected by any propeller blast.

On some of the older airplanes, the ailerons used to "go out" altogether as the ship approached the stall, that is, they became completely ineffective, and the stick moved right and left freely, as if the control cables had come disconnected. This was an unmistakable stall warning, not likely to go unnoticed even if the pilot was under stress. But at the same time it was also a serious danger because it meant loss of lateral control in the stall. On modern airplanes, the designer tries to give the pilot some lateral control even in the stall; this makes the airplane much safer in the stalled and almost-stalled condition but it also means that the ailerons do not "go out" and hence no longer give quite so clear a stall warning.

SENSING ANGLE OF ATTACK

The pilot can gauge the buoyancy of his airplane also in more direct fashion. In the last analysis "buoyancy" or "lift" is Lowness of Angle of Attack: and there are ways of sensing the Angle of Attack direct. At different Angles of Attack, the air flow on the airplane is of course entirely different. At low Angle of Attack, the wind blows from the direction in which the nose points. At high Angle of Attack the wind comes up against the airplane more from underneath. This different airflow produces all sorts of different clues. The changes in the sounds of flight, already discussed, are partly due to this; not only the *speed* of the airflow changes as the speed changes but *direction* of the airflow changes also. Various burbles, eddies, and vortexes will develop somewhere on the airplane at one Angle of Attack and disappear at another Angle of Attack.

Because the direction of the air flow differs in various flight conditions, the pilot can sometimes actually smell the approach of a stall! At high Angle of Attack (slow speed, low buoyancy) smells and hot air from the engine may be wafted up to the pilot's head. Or a different air circulation may set up within the cabin so that the pilot smells gasoline, or bits of dust picked up from the cabin floor.

Here are more clues, all due to the different *direction* of the air flow when the ship flies at high Angle of Attack. In some ships the air strikes the tail in slow flight at such an angle as to set up a high-frequency vibration which the pilot can feel in the stick—not a rough buffeting that goes with the actual stall, but something that feels to your fingertips as if you were touching a vibrating musical instrument.

HARD OR SOFT RIDE

An airplane's "ride" differs at different degrees of buoyancy, that is, different speeds and different Angles of Attack. Quite apart from the different feel of the controls and the different response to the controls, there is a different feel to the whole airplane itself, which even the occupant of a passenger seat can sense, provided he is a pilot, that is, provided that he is awake to that sort of thing.

As usual with such clues, a student can best catch on to it by observing extreme cases first. A really fast airplane, flown through ordinary rough air, gives you a continual hard spanking from underneath, and once in a while it tugs you down sharply by your safety belt. But fly that same airplane slowed up to nearly stalling speed through the same rough air, and you get still some sinks and rises; but the sinks are gentle and the rises have no force to them. That's why, in really rough air, a pilot of the usual civilian ship must slow down; at ordinary cruising speed the sudden sharp increases in lift, due to upgusts, might put excessive stresses on the wings.

That the same airplane, flown through the same rough air more slowly, should be so much less affected by its roughness is due to several aerodynamic-mathematical reasons; but rather than to explain the phenomenon, our problem here is to show how it provides the pilot with a clue to "lift," that is, buoyancy. Reduce this effect to very small scale, and you have that clue—the different ride of an airplane at slow and fast speeds, high and low angles of attack. The air is almost never perfectly smooth. As an airplane traverses 2 or 3 miles a minute, it cuts through thousands of small bits of turbulence. And it behaves with respect to that small-scale turbulence just as it behaves with respect to the bigger bumps; it has a rough, hard, vibrant ride at high speed and low Angle of Attack, a soft, dead-feeling ride at low speed and high Angle of Attack.

That a pilot should be able to perceive such small differences may seem surprising. But it is no more remarkable than that you can sense a difference between the types of road surface over which you drive a car. The car has rubber tires, and springs, and shock absorbers, and upholstered seats, and yet you sense through all that the different feels of macadam, concrete, or gravel. In the airplane, you sit practically uncushioned on the wings. But it is true that in heavily wing-loaded airplanes this effect is less marked than in more lightly wing-loaded ones—for the same aerodynamic-mathematical reasons that make the more heavily wing-loaded of two airplanes—both flying at the same speed—less affected by gusts and bumps. This seems to agree with the observation of pilots that such ships have relatively little "feel" to them and that they force the pilot to rely more heavily on his air-speed indicator. And this, in turn, would seem to prove that the ride of the airplane *is* an important clue to speed and buoyancy; that, even though the pilot may not think about it, he does feel it and it does help him judge.

STICK PRESSURE

Another clue to Angle of Attack—and hence to buoyancy—is the noseheaviness of the airplane, as felt in the stick: how much back pressure on the stick is required to keep the nose of the airplane from going down. As explained elsewhere, the horizontal stabilizer tends to keep the airplane always at one certain Angle of Attack; usually the low Angle of Attack that corresponds to cruising speed. If the airplane is to fly at any higher Angle of Attack (any nearer the stall), it must be forced to that angle by bringing the stick back, and must be *held* at the new angle by *holding* the stick back, with continual back pressure. Hence a properly behaved airplane will never *slow up* and will never *fly* at high Angle of Attack. It must *be* slowed up by the pilot; it must *be* flown at high Angle of Attack by the pilot, by continual back pressure on the stick. The higher the airplane's Angle of Attack the more noseheavy will it feel to the pilot, and the greater will be the back pressure needed on the stick to keep it from nosing down, picking up speed, and resuming its original low Angle of Attack.

Many pilots would call this the most important single clue to "lift." But it can be slightly deceptive. The trim tab can falsify it. When the pilot sets the trim tab control to "nose up," what he really does thereby is to keep his flippers deflected upward without having to hold back pressure on the stick with his own hand. The airplane is then "really" still noseheavy, but the pilot won't feel it. It is at high Angle of Attack and near the stall—but the stick feels as if it were cruising. It will then take only quite slight manual back pressure on the stick to force the airplane to dangerously high Angle of Attack—and a stall. Thus noseheaviness is a buoyancy clue only if a pilot remembers how the trim tab is set at the moment—or if he can check by a quick glance at the trim-tab position indicator which is provided on most airplanes.

BIG SHIPS HAVE NO "FEEL"

And another caution should be added: a big, powerful airplane usually becomes unstable when slowed up and brought to high Angle of Attack with power on; that is, such a ship may then *not* want to return to lower Angle of Attack and faster flight, but may want to slow itself up and stall itself. It may *not* become nose-heavy, but may actually become tail-heavy! Hence the pilot, holding the ship in the slow-flight, high Angle of Attack, low-buoyancy condition, may not necessarily be holding back pressure on the stick: the stick may be forceless, or may even require forward pressure! This makes stick pressures quite useless as a means of gauging the flight condition of such a ship, and that is perhaps the main reason why such ships are harder to fly well, and almost impossible to fly well without heavy reliance on the air-speed indicator.

BEWARE OF THE CREEPING HAND

And here is another caution concerning the noseheaviness of the airplane, used as a clue to buoyancy. Suppose that a pilot is making a gliding approach. His attention is mostly on his landing field and perhaps on traffic, and meanwhile he gauges his gliding speed and buoyancy reserve mostly by the ship's noseheaviness, that is, by the amount of back pressure required to keep the nose from going down. This back pressure seems to him just about constant, and that leads him to believe that he is holding a nice steady safe gliding speed. Actually it is quite possible that he is gradually slowing himself up toward a stall.

How that can happen is best explained by an "exploded" account. In the very beginning, when the throttle is cut and the glide begun, the ship's nose is pointed a little too high. As the speed slackens, it wants to go down. The pilot, trying to hold everything steady, responds to this by bringing the stick farther back. This keeps the nose up but means a further loss of speed, and presently the nose wants to go down again. The pilot again brings the stick farther back and succeeds once more in holding the nose up—at the cost of a further loss of speed; and so forth. Imagine this process not as a sequence of rough jerky stages but as a continual smooth blending of causes and effects, and you can see how a pilot, while holding constant back *pressure*, can nevertheless pull the stick gradually too far back and pull himself into a stall.

It may be objected that this can't really happen: the pilot would notice, as his hand comes back, that the stick force required becomes greater and greater.

A *well-behaved* airplane becomes more and more noseheavy as the Angle of Attack increases and thus requires heavier and heavier back pressure on the stick. Thus a *steady* amount of back pressure can't cause the stick to creep all the way back; it would take a steadily *increasing* amount of back pressure to do that. But all ships are not well behaved in this respect. There are ships in which your hand can creep all the way back rather easily without uncovering much *additional* resistance as it comes back. There are even small ships whose sticks become *light* toward the end, just before the stall. Also, the pilot's sensing of the amount of back pressure he is using is not at all reliable —his arm tires during a long approach glide; he takes a new hold on the stick once or twice. Add to this the natural tendency of the pilot to come back on the stick anyway in an emergency situation, as the unwelcome ground approaches, and you can see that noseheaviness alone is not always a reliable clue to buoyancy.

When gliding slowly near the ground, therefore, and especially in a tense situation, the pilot will be wise to arrest his hand consciously once in a while to make sure that he is not allowing it to creep back. If the ship thereupon noses down, that is a sign that he was gradually pulling himself into a stall, in the manner described; the remedy is to hold the stick steady and let the nose seek its own level.

STICK POSITION

The most correct way, in the engineering sense, of gauging one's buoyancy is by the *position*, rather than the *feel*, of the stick. Flight instructors won't like this statement. At present it is considered bad form to speak or even to think of control *positions* or control motions. You are supposed to think only of the *pressures* you exert on the controls—because such thinking makes for a better control touch, a smoother control action. But the fact remains that stick position is a fairly precise indicator of buoyancy, at least in simple, well-behaved airplanes such as primary trainers or private "family" ships.

Here is why: "Buoyancy," remember, is nothing but "Lowness of Angle of Attack." But the airplane's Angle of Attack control is the "elevator." With the stick in any given position in the fore and aft sense, the well-behaved airplane will promptly assume a certain Angle of Attack; and it will then stay at that certain Angle of Attack as long as the stick remains in that position. The farther back the stick, the higher the airplane's Angle of Attack. Whether the airplane goes up or flies level or goes down at that Angle of Attack depends not, as most beginning students think, upon the stick position, but entirely on the throttle setting. Thus stick position, since it *causes* Angle of Attack, also *indicates* Angle of Attack and thus indicates what the pilot wants to know—his buoyancy or "lift," that is, how far he is from the stall.

For example, a pilot may want to know during a very slow glide how far he is from the stall. Now a stall is nothing but a flight condition of excessive Angle of Attack. However crazy an airplane's *attitude* may be, it cannot stall unless its wings are meeting the air at that certain excessive Angle of Attack. The wings won't be at stalling Angle of Attack unless the flippers, back on the tail, are deflected upward far enough to force the airplane to that Angle of Attack and hold it there. Hence, unless the pilot brings his stick to a certain back position in the cockpit and holds it there his airplane *cannot stall!* Exactly what that position is in any particular airplane it is very much the pilot's business to find out when first "checking out" on that airplane. In most types this stalling position of the stick is quite far back, quite near the pilot's stomach; hence the pilot can judge how far he is from the stall simply by noting the distance between his hand and his stomach.

In a well-behaved airplane, one could actually label the various stick positions, fore and aft, in terms of buoyancy, "lift," or whatever the pilot may choose to call it. That certain back position would be labeled "stall." Another position, a few inches farther forward, could be labeled "very mushy." Another few inches farther forward could be the position "normal glide, with good healthy lift—also a good healthy climb." Still another few inches farther forward, another stick position could be labeled "cruising, very firm sustentation, lots of lift." Still farther forward would be a position "very fast flight, sustentation so firm as to be actually hard; when in rough air watch out for structure of wings."*

There is a catch, however, to stick position as an indicator of buoyancy. There is some catch to any one clue that a pilot could possibly use! In this case, the trouble is that the pilot has no good way of gauging the position of his stick. He is much more conscious of stick *pressure* than of stick *position*. Unless he actually looks at his hands, he may bring them quite close to his stomach and not realize it—simply through a case of creeping hand, as described earlier. Neither has he a reliable way of remembering where the stick ought to be—in a glide, for instance. He cannot simply move the stick to the desired position, the way you slip an automobile's gear-shift lever into the desired notch, or the way you set the throttle of an Army trainer to certain remembered positions on the quadrant for take-off, for climb or for level flight.

Moreover, in some airplanes, especially in low-wing monoplanes, the range of stick travel is very small. This tends to be especially true with power on. The stick of such a ship may be only a couple of inches farther forward in a fast power dive than it is in cruising flight; and in a very steep climb, done quite near stalling speed and quite "mushy," the stick may be only a couple of inches back of its cruising position. This makes the gauging of stick position much too hard to be of practical use as a flying clue. Training airplanes ought not to have this characteristic; a training airplane should require wide, highly noticeable changes of stick position for small changes in Angle of Attack.

But then, nobody has yet built a training airplane clearly and sharply designed as a trainer. Training airplanes also should have muffled exhausts and quiet propellers, so that instructor and student could talk. At present, we teach and learn one of the most demanding arts by deaf-and-dumb methods!

A training ship should have an indicator that would show, by pointer and dial on the instrument board, in what position the pilot is holding the stick—similar to the indicators which show trim-tab

---

* These statements will seem a little dubious to a reader who thinks that the "elevator" is the airplane's up-and-down control. A later chapter will discuss the functioning of the "elevator" in more detail, and will show that it really is the airplane's Angle of Attack control.

position and similiar to the indicators on a steamship that show rudder angle to the helmsman. Such an indicator could then be calibrated in terms of Angle of Attack, and also in terms of buoyancy or closeness to the stall.

Such a stick-position indicator is suggested here not entirely as a joke. It would be valuable because it would be one more means by which the pilot could keep a check on himself; and it would constantly remind him of the true purpose of the "elevator," so easily forgotten: that it is an Angle-of-Attack control, not an up-and-down control. Test pilots, who really have to know what they are doing, use control position indicators: why not students and instructors?

Another catch to using stick position as an indicator of buoyancy is this: our present airplanes are not entirely well behaved. Their Angle of Attack is *controlled*, true enough, by stick position; but it is haphazardly *influenced* also by throttle setting and (to a lesser degree) by the loading condition of the airplane.

With power on, a given back position of the stick will in most airplanes produce a higher Angle of Attack and hence a lesser degree of buoyancy, slower flight, and greater nearness to the stall, than with power off. Most of our civilian airplanes will not stall with power off unless the pilot holds the stick clear back against his stomach; with power on many of them will stall when the stick is held only halfway back. This is because the propeller blast hits the upward-deflected flippers and makes them unduly powerful; and in some airplanes because the propeller thrust, pulling forward low on the airplane coupled with the drag, pulling backward higher up on the airplane, tends to force the airplane to higher Angle of Attack; and for other reasons. It is not so in all airplanes; but where it is so, it rather devalues stick position as a clue, since the pilot has to remember a different critical position for each different throttle setting.

This confusion is especially confusing during flight with reduced power—say a power-on landing approach; in that condition, neither the sounds of flight nor the looks of it are much of a guide to buoyancy and speed, and stick position is also tricky—even a little power makes the stick unduly effective in stalling the airplane. But during a power-off glide, stick position *is* a reliable guide, and that is something to remember during a forced-landing approach: *if the pilot's hand is near his stomach, the airplane is near the stall, however it may feel, sound, or look.*

TESTING THE CUSHION

Accurate sensing of one's buoyancy is especially important during a landing approach; particularly so if the approach is to a small field. The pilot then wants just exactly the right amount of buoyancy. He wants just enough to have a slight cushion, so that, when he arrives at ground level, he will be able to flare out, check his descent, and perhaps float for a moment before touching the ground. Any excess buoyancy beyond this would mean a long "float," using up too much room. Any less buoyancy would mean a stall. Pilots call this "flying on edge." And the problem is, of course, to get to the edge and still not fall over the cliff. How can the pilot judge?

In addition to the clues already discussed, here is one that works especially well in this particular situation. The procedure might be called *testing the cushion*. What the pilot wants to know, in the last analysis, is whether he has a cushion of reserve lift, whether the final pull-back on the stick will actually result in a surge of lift and thus a firm checking of the descent, or whether it will result in a stall and a drop. Well, the best way to find out is to try right then and there: pull back on the stick and watch how the airplane behaves.

Take some extreme cases first. Suppose that the pilot is in an extremely slow "mushing" glide, at the very edge of the stall. In that flight condition, when the stick is moved back a couple of inches, the flight path does not go up; the ship begins to stall and settle and the flight path goes down. But if the pilot makes the same backward motion with the stick during an extremely fast glide, when he has lots of buoyancy and very firm sustentation, he gets a lively upward deflection of the flight path.

The pilot is, of course, too high above the ground to *see* this upward deflection, or the absence of it (the way he can during an actual landing flare-out, when the nearness of the ground makes the slightest ballooning or the slightest settling clearly apparent to the eye). He must judge not by eye, but by kinesthesia. He notes the small changes in his own weight, the feelings of lightness or heaviness that result from his handling of the stick. If the glide was brisk and the small pull-back on the stick makes the ship come up against him from underneath, he feels slightly heavier for a moment—the way one feels when an elevator starts upward. If the glide was "mushy" and the ship fails to come up against the pilot, he feels no change and the ship seems to have "not much lift"—feels "dead." If the glide was extremely slow so that the pull-back on the stick makes it settle out from under the pilot, he feels slightly lighter for a moment—the way one feels when an elevator starts downward—and the ship feels "gone."

Those are the extremes. Between them the pilot can distinguish many graduations. One's perception of one's own apparent weight is quite sensitive—provided, of course, that one's attention is directed to it. And a large part of the art of flying consists largely in paying attention to the sometimes rather odd things that matter!

Many pilots "feel" for their "lift" in this fashion almost continuously during an approach—especially during the last stage where the approach glide blends into the actual landing, and during night landings. If you watch such a pilot, you see him every few seconds gently tugging back on the stick an inch or so, and then letting it come forward again—quite a different thing, of course, from the hasty stick pumping of an overcontrolling beginner. If you ask such a pilot what he is doing and why, he will perhaps say he is "hefting" the stick and feeling out the stick pressure, the noseheaviness of the ship. But what he is really hefting is his own body: he is feeling for the changes in his own heaviness caused by those small motions of his hand. This is largely what a pilot means when he claims that he can feel the "lift" of his airplane in the seat of his pants.

SEEING THE LIFT

Toward the last of the approach, quite near the ground, this same clue works also via the eyes. If a small pull-back on the stick does not result in the visible ballooning, or at least in a visible checking of the ship's descent, the ship is on the very verge of a stall. This is the clue a skillful pilot goes by when making an extremely short landing in very restricted space. He gets well away from the intended landing field and makes his approach low and shallow with power partly on. As he gets near the ground—say within 30 feet of the ground—he coaxes the ship into extremely slow nose-high flight; that is, he flies it at almost stalling Angle of Attack, with almost no buoyancy. At first glance this would seem dangerous so close to the ground, and for an unskilled pilot it is very dangerous indeed; he might easily lose that last bit of buoyancy and stall in, hitting the ground nose-first. But for the skillful pilot, this very closeness of the ground makes the gauging of his buoyancy extremely accurate, and allows him to fly closer to the stall than he would dare at slightly higher altitude. The working of the controls during such an approach will be discussed elsewhere in this book. What matters here is that this method brings him over the beginning of the runway with neither excess altitude nor excess buoyancy. He then cuts his throttle, and the ship squats then and there.

CAN "FEEL" BE PRACTICED?

This about finishes our list of clues to "lift" in flying. It may be well to make quite clear that these are *not* clues by which, in the writer's opinion, a pilot *ought to* judge his "lift." They are the clues by which—according to the writer's observation—the pilot *does* judge his "lift."

"And so—what?" The reader may feel that this listing of speed and "lift" clues is somewhat academic. "All right, maybe these are the clues that pilots go by; it follows then that after completing flight instruction I, too, shall derive from them a sense for speed and lift." But theoretical statement for the sake of statement is not the purpose of this book. The writer believes that the process of learning to fly could be quicker and surer if students would know more clearly what to watch for, if instructors could point out more precisely just what it is the student is supposed to learn. Sometimes the student needs a clearer idea of the mechanics of flight, for example, as regards the Angle of Attack, and the exact nature and function of each control. Sometimes he needs a clearer understanding of his own functioning; and that is what this chapter is to contribute.

Look at it this way: We demand from the prospective student flier that he have perfect eyes, perfect sense perception in all respects. And we demand also a great natural aptitude in coordinating perception and action, an aptitude that he must prove by various tests. And indeed it can't be denied that half the art of flying consists of perception. But once we start the student on his training, we don't teach him how to use his fine senses; we leave that to chance.

It is suggested, then, that the student try out each of these clues separately, and try it out in a sharp, exaggerated fashion. With the instructor in the ship, try sometime to maintain a steady gliding speed by ear alone, with your eyes closed. Or let the instructor glide the ship and try to guess the gliding speed in miles per hour with your eyes closed. Try once or twice to judge the glide quite mechanically by stick position; upon closing the throttle indicate with one hand the location to which you are going to bring your stick hand so as to get a certain gliding speed you want and see what speed you get. Sometime, try hefting the stick during an extremely fast dive; then again during a very mushy glide.

Remember, there is a great difference between merely *perceiving* something and *noticing* it. A savage, put on an American city street, would see the traffic lights just as you do—maybe better. He would probably overlook them and watch instead the flashing neon signs, the lights of cars, all sorts of other clues that are more impressive but much less important; for he would not know what a traffic light means. But we see traffic signs even with bad eyes and while thinking of something else because we watch for them and understand their meaning instantly and know that, though they are not very attention-catching, they are important.

When the flight surgeons test our eyes, ears, balance, and so forth, they are worrying unnecessarily—certainly as concerns civilian flying. A pilot needs no better vision than does a bank clerk or a housewife, nor does he need better hearing, balance, depth perception, and all the rest. The clues by which we fly, the things our senses must pick up in order to enable us to fly, are all perfectly plain, and can be easily and plainly perceived by eyes, ears, and so forth, of less than average keenness. Our difficulty in learning to fly is not sense perception, but interpretation of what our senses perceive. We tend to pay attention to the wrong things; we miss the things that matter because we aren't looking for them, because we do not know what they mean.

We notice those things that matter to us; a mother hears her baby's crying in a distant part of the house right through the chatter of a dinner party. The same thing goes for the dumb student flier—show him what the signs are and why they are important, and he won't be so dumb. Once his attention has dwelled on them a few times they become much more noticeable; once the correct response to them has been practiced formally a few times, it becomes almost automatic. And that is all the so-called "flying instinct" consists of: small clues, understood correctly and reacted to automatically.

THE AIR-SPEED INDICATOR

But what about the air-speed indicator? Can't you always tell by the air-speed indicator how far you are from the stall, how much buoyancy you have?

The answer is neither a straight yes nor an outright no. The air-speed indicator is indeed the pilot's most important flight instrument. As our airplanes are now equipped, it is the only instrument that indicates anything at all concerning buoyancy, Angle of Attack, closeness to the stall. But it is not a simple and straightforward instrument. In the first place, it can develop mechanical trouble (and, one is tempted to add, it usually does). In the second place, it has some peculiarities that must be understood before it can be used as a buoyancy meter or a stall-warning device. Ignorance of its peculiarities has cost many a pilot his life.

First of all, it is misnamed; it is not a speed indicator at all. It is actually a pressure gauge. It measures the dynamic pressure built up by the impact of the air upon the airplane as the airplane advances against the air. It ought to read not in miles per hour but in pounds per square foot, like a tire pressure gauge, or in inches of mercury, like a barometer. Since the impact pressure increases when the airplane advances faster and decreases when the airplane advances more slowly, it is possible to calibrate this pressure gauge in terms of air speed. But such a calibration can be correct only for air of a certain density— sea-level air of moderate temperature. At higher altitudes, when the airplane advances against the air at the same speed, the air's dynamic pressure will be less because the air is less dense (has less weight per cubic foot). Hence the air-speed indicator will underindicate. The same thing is true at sea level if the air is unusually hot, because hot air is lighter. Hence the air-speed indicator underreads on a hot day. On a very cold day, the air-speed indicator always shows high values— a fact that airplane salesmen know very well!

This makes the instrument somewhat awkward for navigation purposes. The navigator really needs to know how fast he is moving; hence he must always correct his air-speed reading for temperature and altitude. For the pilot, it works differently. The pilot is really interested not in speed, but in stall danger; he wants to know how far he is from the stall. And in that respect the indications of the air-speed dial are *not* falsified by temperature or altitude. If in any particular airplane, near sea level, the air-speed indicator reads 60 m.p.h. just as the stall occurs, then 60 m.p.h., *as shown on the air-speed indicator*, will be the stalling speed of *any* altitude and in air of *any* temperature. The actual stalling speed is much higher at high altitude; at 5,000 feet the stall of that particular airplane will actually come at more nearly 66 m.p.h. At 20,000 feet, it will actually come at about 90 m.p.h.; but regardless of altitude and *actual* speed, the air-speed indicator will always *show* 60 m.p.h. just as the stall occurs.

In this respect, then, the air-speed indicator is an ideal stall-warning device. It is a poor speed indicator but a good indicator of one's buoyancy. Regardless of altitude or air temperature, if the instrument shows a margin above stalling speed, you have that margin of buoyancy. If the instrument shows no margin, you have no margin.

But in another respect, the air-speed indicator is an extremely poor stall-warning device. The stalling speed of an airplane depends on its weight. Flown empty, with no load and hardly any fuel, a certain airplane may stall at 60 m.p.h. That same airplane, overloaded to a point where it can just barely stagger into the air, will weigh about twice as much. And its stalling speed then will be about 85 m.p.h.!

Thus in a given ship, a certain air-speed reading may mean an ample safety margin, plenty of buoyancy, when the ship is light; with the ship loaded, the same reading may indicate an extremely mushy condition, quite near the stall. This is important not only in avoiding a stall; it is equally important in avoiding unnecessary excess buoyancy and excessive floating when making a landing approach with the ship light and empty. The emptier the ship is of load, the lower is the air-speed figure at which the approach glide should be made.

Training airplanes are not usually flown with widely varying loads. But there is another way in which the weight of an airplane can increase and decrease—by maneuvering. A 60-degree banked turn, for example, has the effect of making the airplane twice as heavy as it is in straight flight. Hence, in such a turn, the stalling speed is almost one and a half times what it is in straight flight! Imagine then the embarrassment of the pilot who "knows" that his airplane's "stalling speed" is 60 m.p.h.—and who then finds himself spinning out of a steep turn while the air-speed indicator reads a supposedly safe 85 m.p.h.

What is true for a turn to the right or left is true also for any pull-out from a dive. Centrifugal force loads the airplane down, and the usual straight-flight stalling speed becomes meaningless.

It is sometimes suggested, in view of all this, that the air-speed indicator should be redesigned. By integrating it with an accelerometer (an instrument that measures the pull of centrifugal force on the airplane in curving flight), one could achieve an instrument that would always truly indicate how close the airplane is to the stall, and at what Angle of Attack it is flying—in short, its buoyancy.

ANGLE OF ATTACK INDICATORS?

Certainly some instrument is needed that would tell the pilot exactly what his Angle of Attack is, that is, how much buoyancy he has, how close he is to the stall. But the art of flying is still in a primitive state. The most important fact about an airplane's flight condition is not indicated by any instrument. This is not because such an indicator cannot be built but because designers don't appreciate the need for one. Perhaps they are right—too many pilots don't know what Angle of Attack is in the first place, and such an instrument's indications would be meaningless to them.

When an Angle of Attack indicator is proposed, aeronautical people usually take it for granted that it would have to serve primarily as a stall-warning indicator. And promptly they shy away from it, for two reasons. They think it inadvisable to relieve the pilot of the responsibility for avoiding a stall; the pilot would then simply keep coming back on the stick until the red light would flash or the horn would blow, or what not. He would lose his healthy respect for the danger of stalling. Also, the experts point out, it is difficult to build a stall-warning device that will indicate accurately under all conditions, including icing, and would not be liable to mechanical failure; and what if the pilot relied on the red light to warn him, and the red light flashed too late?

These arguments make sense, but the original assumption does not. The most important role of an Angle of Attack indicator would not be to give last-second stall warning; it would be to keep the pilot *continually* appraised of his Angle of Attack "lift," buoyancy, or whatever you might call it. Pilots don't stall and spin when they are consciously trying to fly as slowly, as close to the stall, as possible; on such occasions, they are alert. They stall and spin out of turns with power on when usually they have no idea that they are even anywhere near the stall! A stall-warning indicator, therefore, is like a gasoline gauge that would indicate only when the tank is empty, or a bank that would send a statement only when you are overdrawn. What the pilot needs is a device that will warn him *early* that his Angle of Attack is *beginning* to increase. Instead of flashing a light in a 70 degree banked turn, when the whole airplane is all on edge, it should warningly show the first slight increase in Angle of Attack at 30 degrees of bank; then show at 45 degrees how the margin of safety has become a little less; then show how at about 60 degree angle of bank the safety margin rapidly decreases.

The best use of an Angle of Attack indicator, however, would be not as a safety device for practical flying but as a training device. As it is now, a student can go through a quite elaborate series of primary and advanced courses and never clearly realize the most important fact of all flight—Angle of Attack. He may not even know that there is such a thing! He certainly does not know how it works out in climbs, glides, turns, slow flight, and fast flight; how it changes in gusts; and during maneuvers how it reacts to changes of stick position and throttle setting. And thus he misses the central idea of his whole art.

The accident record is one result. Another result is the rate of "wash-outs" from flight training. The worst result is the enormous number of people who never even begin to fly because they are not supposed to be fit to fly. It is absurd that 40 years after Kitty Hawk, still only a tiny percentage of all human beings is considered fit to fly a really businesslike airplane of high performance. But it is not surprising. Flying an airplane when you don't understand Angle of Attack is indeed an art of the very highest order. With the main idea veiled in secrecy, of course the pilot has to go by hunches and feels, make quick decisions, act against his own instincts, supress all sorts of fears. Of course, under such circumstances, he needs superior perceptions, an exceptionally stable personality, great presence of mind, and all the rest. Of course the airplane appears to him a tricky demon whose behavior he can't ever quite predict, a dangerous contraption that often refuses to obey the controls and goes spinning down if you pull too hard on the 'up" lever.

But with the idea of Angle of Attack once clearly understood, flying is simple, common sense, logical, not trying to the nerves; it is easy.

The Wright brothers knew all this. Their approach to the art of flying was much more sophisticated than ours is today. And the only flight instrument they ever designed was an Angle of Attack indicator! It was merely a piece of string tied to the nose of their airplane, so that its streaming would indicate the direction of the Relative Wind. If it streamed back level, the Relative Wind was from level in front, and the airplane was at low Angle of Attack. If it fluttered upward, then they knew that the Relative Wind blew at the airplane from underneath, that the Angle of Attack was high and the airplane near the stall. The same tuft also indicated skid or slip, that is, sideways motion through the air. The device was, of course, the same simple thing as the pennant on the masthead of a sailboat, whose streaming indicates the direction of the boat's "apparent" wind and by which the sailor trims his sails. On today's airplane, we could not put such a tuft on the nose of the airplane, since the propeller blast would make it give false indications. Perhaps it should be carried on a pole extending forward from one wing, well as the Pitot tube of the air-speed indicator is carried now—only perhaps farther forward, so that it would be in undisturbed air and well in the pilot's field of vision. Perhaps every flying school ought to have at least one airplane fitted with such a forward mast and pennant, and every student should perhaps have a chance to fly it once in a while and see what Angle of Attack really means.

PART II

SOME AIR SENSE

Before you look deeper into the question of how an airplane is controlled and maneuvered, it will pay you to think through in a general way just what it means to *fly:* to go up into the air and lose your connection with the solid ground. You will meet there with some quite unexpected effects.

Some of them are difficulties. For example, the airplane can be steered in the up-and-down sense as well as the right-and-left sense. The up-and-down steering itself is simple enough; but it brings on certain complications that will keep bothering you in flight, will in fact make a fool out of you, unless you have once thought them through. Another example is the effects of wind upon the airplane. They, too, must be thought through. To do so requires some mental effort. But it would require much greater effort—not to mention embarrassment, harassment, and expense—to try, as most student pilots want to try, to worry these things out in flight.

Some of these unexpected effects are not difficulties but helps to the student pilot. For example, the tendency of the airplane to fly itself. Most beginners think that it is literally the pilot who does the "flying," that he keeps the airplane up by a mysterious knack, a sort of a balancing trick. And they think that the moment the pilot stops "flying" the airplane, or makes a mistake in his "flying," the airplane will fall down. Actually the airplane largely flies itself and with but little exaggeration one could say: never mind how the airplane's controls work. The most important thing to know about the controls is that the less you use them, the more gently and lazily you use them, the better the airplane will fly. Most of the time, the airplane flies not *because* of the pilot's activity on the controls, but *despite* it! And if you ask yourself what piloting really consists of, the first and most important answer is: it consists 90 per cent of doing nothing at all!

You would find that out even if it were not explained to you. But here again you will learn faster if you have thought this effect through beforehand, and if you understand it. More important, you will be more relaxed as you learn.

Chapter 5

THE LAW OF THE ROLLER COASTER

HERE is a thing that used to happen often, back in the days when flight instructors didn't know so much about the instructing side of their business. During a glide the instructor would say, "You are gliding too fast. Slow it up a little." The student would not respond. The instructor would say, "Come on, can't you hear me? I said slow it up." The student would think, "What the devil does he want me to do? My throttle is back as far as it will go, and this contraption hasn't got any brakes. How can I slow it up?" And so he would do nothing at all, and presently the instructor would get mad and grab the stick—

It was just one of those typical misunderstandings between flier and nonflier. Some things about flying have become so obvious to any pilot that they go without saying—and therefore he sometimes forgets to say them. The beginning student, on the other hand, is disoriented; he sometimes can't see the forest for the trees. He sometimes can't see the obvious things just because they are obvious—when he is all concentrated to learn something fancy, something intricate, something requiring skill.

What the instructor wanted the student to do was simply to hold the stick a little farther back and point this ship's nose not quite so steeply down. And what he might have said was, "You are gliding too steeply. That way, you pick up too much speed. Don't let your nose point down so steep, and you won't go so fast. This thing works pretty much like a sled."

Like a sled, an airplane with the engine throttled keeps going by coasting down a slope—sacrificing height in order to maintain speed. The only difference is that in an airplane you can pick, literally out of thin air, any slope your heart may desire. You can slide down a steep slope and go like the very devil, or you **can choose** a shallow slope and glide more slowly—and longer. The **normal** glide of an airplane is rather shallow, with a speed very much less than level cruising speed.

But what the airplane "wants" to do with the engine throttled back—that is, what it will do if you take your hand entirely off the stick and give it its head, is to glide very steep and fast—in fact to dive, at a speed far exceeding level cruising speed. To hold an airplane in a normal glide therefore requires a constant back pressure on the stick so as to keep the nose from pointing down too steeply; and that's what the whole little episode was about. But note how the pilot's "slow it up" meant the same thing as "pull the nose up higher."

SPEED = HEIGHT

This is, however, only the first and most obvious example of a law that affects the whole art of flying. This effect is never discussed in the manuals, and it has no official name. One instructor once, arguing that it is an important part of the theory of flight and that it ought to be formally explained to every student, called it *"those exchanges you have to make—you know."* And with this expression he hit the nail on the head. In an airplane, "slow" and "fast," "high" and "low," "up" and "down," "lift" and "drop" are tied up with each other in a peculiar fashion. You can always get one by paying for it with the other. You can never get rid of one without getting a lot of the other. The comparison with a sled does not cover all of it. The handiest comparison we have is at the amusement park—the roller coaster.

A roller coaster starts out by being high—having been hauled to the top of the track by some mechanical hoist. And it starts out, on the top of the hump, by having very little speed. Then it cuts loose, and in going down it converts altitude into speed; the steeper and farther it goes down, the more speed it picks up. * At the bottom, it has no altitude, but plenty of speed; and it presently proceeds to convert speed back into altitude by shooting up the next incline. At the top, it is almost out of speed; but it has almost all its original altitude again. And then it goes down another slope.

The same thing is true when an airplane glides with the engine throttled back; speed and height are two forms of the same thing. "Of course they are," says the physicist. "They are two different forms of energy." When a pilot maneuvers, he continually makes exchanges,

---

* This is loosely worded on purpose. The precise mechanics of acceleration are more complicated, but, since an understanding of them would not particularly help the pilot, they are disregarded here.

turning altitude into speed or speed into altitude; and he makes those exchanges whether he wants to or not.

The roller-coaster effect works in many different ways. The most obvious is that (with the engine throttled back) you can't get speed without sacrificing altitude, nor can you maintain speed without paying out altitude. If you ever tried to maintain your altitude, you would presently lose your speed; and, once you had lost your speed, you would lose your lift and would stall.

The same roller-coaster law works also the other way round. Once you are stalled, there is only one way to regain speed and lift: point your nose down and dive the airplane; pay out altitude to purchase new speed.

A light airplane might have to sacrifice 75 feet of altitude in exchange for new speed and lift; in a heavy bomber the same thing might take several thousand feet of altitude. In any airplane, the pilot is extremely careful not to slow the airplane up too much unless he has enough air space under him for recovery from a stall—plus a big margin of safety. If he has a lot of altitude, on the other hand, then he may safely stall the airplane for practice or just for fun; and he may safely try some maneuver that might possibly result in a stall, for he can always use his altitude to purchase new speed and lift. Thus, as one writer on the art of flying puts it, "Altitude is money in the bank."

But if altitude is money in the bank, speed is money in the pocket. For just as altitude can be converted into speed, so can speed be converted into altitude—the roller-coaster way. The manufacturers of one fast-cruising airplane once used this idea in an advertisement aimed at pilots. You could hedgehop that ship safely right down at tree level, they claimed, and if your engine quit suddenly, you were perfectly safe—because its tremendous excess speed above stalling speed allowed you simply to point your nose up (converting speed into "lift"); you could zoom and gain 1,000 feet of altitude before your excess speed would have dissipated, and, from that altitude, they argued, you could then look around, pick a suitable field, and make a regular forced landing.

DON'T GO BROKE

The one fellow who is really broke in the air is the one who is out of both altitude and speed. *Low and slow* is the pilot's idea of dangerous flying. *Low and fast* is fairly safe if you don't get to daydreaming and hit a tree, and if you don't let them catch you at it, for it is illegal. *High ana slow* is fairly safe if you do it right and if you have trained yourself to react to a stall in such a manner that a prompt recovery will result instead of a spin. *High and fast*, which your girl friend thinks must be awfully dangerous, is the safest. Thus, if you want to keep well, you have to keep speed or altitude, or best of all, some of each.

Sometimes, the roller-coaster law works the other way round. An airplane has no brakes, and the only way in which you can quickly get rid of excessive speed in an airplane is by converting it into "lift" or altitude. The most annoying example of this is the landing. The actual ground contact in a three-point landing can take place only if the ship is flying at very slow speed—stalling speed or nearly so; landing is one time when you want to be broke! The pilot who has glided too fast and arrives over the runway with too much speed suffers an *embarras de richesses;* he cannot land! To get rid of his excess speed, he must put the stick back and point the nose higher—but as he does so his speed converts itself into "lift." If he tries to hurry the slowing up by pulling his nose up too hard (the way a beginning student will sometimes do), the airplane gets so much lift that it actually zooms away from the ground—and makes its "landing" finally 30 feet in the air, with sad results. But even if the pilot pulls his nose up at just the right tempo he is in trouble. All that speed, converting itself into lift, will keep the airplane floating and floating sometimes clear across the landing field; so that, when it finally settles to the ground, all the runway is already used up, and there is no room in which to roll, nothing ahead but fence.

This will annoy you also in another way. It means, that, during the approach to the landing, the speed must be carefully kept moderate, so as not to arrive at ground level with too much speed; and that means that, if you should find that you have made your approach a little too high, you cannot simply nose down, as common sense would suggest. By nosing down you would get rid of altitude all right, but instead you would pick up a lot of speed. This speed would then cause you to float forever in the landing, and you would actually land at just about the same place that you would have reached had you not nosed down but simply continued your too-high approach. This makes it important to learn certain tricks by which you can kill altitude without picking up speed; mushing, sideslipping, essing; some pilots also use certain tricks by which they can kill speed without picking up altitude—"fishtailing" for instance. And it makes it even more important to learn how to judge an approach so that no height-killing maneuvers will be necessary. Some of these techniques are explained elsewhere in this book.

WHERE IS THE BRAKE?

One might ask, why no brakes? That is one of those dumb questions that embarrass the expert because he has no good answer. Indeed, why no brakes? If the airplane had brakes, it would be a great deal easier to handle; if you want to get it down, you would then simply nose it down, putting on the brakes at the same time to keep from picking up speed. You could make your landing approach at ample speed and hence with great safety; when you finally got ready to land, you could simply put on the brakes, kill your speed in a hurry, and sit down.

Brakes are difficult to design, however, and expensive to build, and they would add weight. At present only dive bombers and some very clean sailplanes have air brakes—dive bombers because they must dive nearly vertically and would build up unmanageable speed; sailplanes because they are so exceedingly clean that even the slightest dive would build up speed that might endanger the ship's structure.

In lieu of brakes, many airplanes have flaps. Flaps are sometimes described as air brakes, but that is only half correct. It is true that, once the flaps are down, an airplane has very much more drag than with the flaps up, and hence it can be glided much more steeply without picking up speed. Also, if after a steep fast glide a flare-out is made for the landing, a heavily flapped airplane loses its speed very fast. Hence, once the flaps are down, the airplane can be handled in the approach glide almost as if the roller-coaster law did not exist. If you are a little too high, you can afford just simply to point it down more. It will then pick up additional speed, to be sure; but against the drag of the flaps it can't pick up very much additional speed; and the pilot knows that it will quickly lose that speed in the end, when he flares out.

But in another way, the flaps themselves are subject to the roller-coaster law. You put them on because you want their braking effect. But the first effect of putting them on is a temporary increase in lift which balloons the airplane upward anywhere from 20 to 200 feet depending on the type of airplane, the type of flap, and the speed at which the pilot puts the flaps on. Thus, temporarily at least, speed is converted into "lift" and altitude. Again, you might wish to retract the flaps, perhaps because you have discovered that your glide is too steep, that you will come down short of your intended landing spot, and that you don't want the flaps' braking effect any longer. Yet the first effect of "spilling" them, that is, retracting them, is to cause a temporary loss of lift, which causes a sink—possibly of several hundred feet. If the pilot in that situation did spill his flaps taking off these so-called "air brakes," he would sink right into the ground! Thus flaps, too, are operating somewhat under the roller-coaster law; they, too, convert speed into lift and altitude, and they, too, make it impossible to pick up speed again without first losing considerable altitude.

ENGINES ARE PUNY

But, one might ask, what about the engine? With power off it may be true that you can't speed up an airplane without sacrificing height. But surely you can always speed it up by ramming the throttle wide open? Unfortunately, this is not practically feasible. All that has been discussed for power-off flight is essentially true also for flight with power on.

Of course, with power on you can maintain your height without killing your speed, and you can maintain speed without sacrificing height. But that is just about the only thing the engine can do. For an airplane's power plant exerts not nearly so strong a pull as its loud noises might lead you to believe. Suppose you had a 2,000-pound airplane: the biggest pull that its power plant can possibly exert is only about 400 pounds—a quite puny force.* It is enough force to overcome the friction and resistance of the air and thus keep the airplane going without the help of gravity, that is, to allow the airplane to fly level. Beyond that, it can exert just enough additional pull to pull the airplane up a gentle grade, but only slowly, gradually. Point any airplane, even the most powerful pursuit, too steeply up in the sky,

---

* You may wonder, well, what about all that horsepower—where does all that horsepower go? The answer is that all that horsepower consists of the power plant's ability to exert this pull, puny as it is, at comparatively high speed—perhaps 120 m.p.h.

and it will very soon lose its speed and stall. There is not literally such a thing as an airplane's "hanging on its prop." An airplane's power plant could be designed so that it could pull the airplane straight up— it would have to have a huge, slow-turning propeller. But such an airplane would then not need any wings. It would not be an airplane at all but a helicopter!

Thus, even with power on, you are still subject to the roller-coaster law. If you try to climb too steeply—try, in other words, to pick up too much altitude, you will thereby kill your speed. In "blind" flying, for example, the pilot judges whether his nose is up or down or level entirely by the way his air-speed indicator behaves! But in ordinary flying, too, this effect is important. If you should ever be almost stalled and want new speed in a hurry, your throttle won't give it to you fast enough. The only thing that will give it to you fast is to nose down and let gravity pull you roller-coaster fashion. But that means sacrifice of altitude.

HOLDING IT DOWN

The roller-coaster law is important also in the take-off. You know how right after breaking free a pilot usually prefers to drop his nose a bit and shoot along level for a second or two before he starts his climb. Why does he do that? The engine is giving him so much energy every second. By the roller-coaster law he can take that energy either in the form of additional speed or additional height—but he can't have both. Altitude is money in the bank. Speed is money in the pocket—and the first thing he wants is to have a little money in the pocket before he starts worrying about his bank account. And while he builds up excess speed, he knows that he could, if he chose, convert that excess speed into altitude at any time simply by pointing the nose up.

If the take-off is from a small field surrounded by tall obstructions, the working of the roller-coaster effect can be quite dramatic—if an experienced pilot is at the controls. The inexperienced pilot has a strong tendency to point the nose up steeply and simply hope that the airplane will climb out. But it is easy to overdo this and, by trying to get too much altitude too quickly, to kill one's speed and stall, perhaps 50 feet off the ground. The experienced pilot will therefore point his nose up only as high as absolutely necessary to clear obstructions. In fact, if the take-off is from a very tight field, he will often point the ship's nose actually at the obstructions—maybe a half height of the trees. This does not help the airplane gain height, but it does help it gain speed. And the experienced pilot does not worry, because he knows (without thinking about it) that speed and height are two forms of the same thing, and for various reasons he prefers to have that thing in the form of speed. And he knows that he can always at the last moment convert speed into height by pulling the nose up; the airplane will then "zoom" and clear the obstructions.

Such a zoom may take most of the ship's excess speed away from it—so that it arrives over the treetops all slowed up and nearly stalled. The inexperienced pilot in this situation will hang on, with the stick well back, and wait for his engine gradually to give him new speed; for he is afraid of the ground, afraid to lose any precious altitude. The experienced pilot will not try to "stall" along in this fashion. He knows that it would take too long to get new speed from his engine and that it would thus mean a period of flying slow and low. He knows, too, that the gustiness of the air near the ground might easily cause a stall; and, with not enough altitude underneath the airplane for recovery, that might mean a disaster. What the experienced pilot does, therefore, is to nose down a bit, once he is past the trees, and get back to reasonable speed by sacrificing a little of his height, even though that may mean going down into the next field near the ground once more. So that the onlooker would see the airplane apparently charging straight at the trees, then doing a jump, and then dipping down beyond the trees and disappearing from sight; only to reappear a quarter minute later, now in a healthy gradual climb.

All of which is pretty obvious if you just think it over once.

Chapter 6

WIND DRIFT

A ND then, there is the wind.

Much of the art of flying has turned out much simpler than you had expected: the airplane stable, the air without pockets, the height not at all terrifying. But then, there is the wind: the air, the medium in which you move, is itself in motion. And that now brings on a whole string of quite unexpected complications. You discover that, whatever wind there is, even the slightest breeze, has its effect on you every single minute of flight: it distorts your curves, it falsifies your climbs and glides, it pulls your figure eights out of shape, it makes your ship go one way while its nose points another way. If the wind is across your direction of flight, it makes the airplane sidle over the scenery in a crablike gait: sometimes so much so that, if you want to look where you are going, you must look out the side window of the cabin!

If the wind is on your tail, it makes the airplane hurry amazingly and keeps messing you up by getting you there too soon, before you have had time to plan your next maneuver. If the wind is from ahead, you lose speed; and with a slow ship in a strong wind it may happen that you stand still in the air or even go backward.

It is dismaying to find that mere wind—so flimsy, so intangible a thing—can blow a heavy powerful machine about at will; but it is even more dismaying to discover, by and by, that the effects of wind are almost exactly what your common sense would *not* expect them to be.

PUZZLES

There are endless examples of this. Does an airplane get more lift when flying "against" the wind than when flying "with" the wind? The beginning student will almost always say that of course it does; why, it stands to reason! The experienced pilot says that of course it does not. Again—does a tail wind cause a loss of lift? In any batch of cadets, there will be some who argue yes: the tail wind, blowing against the wing from behind, tends to blow the wings down instead of up! The experienced pilot says no, the tail wind causes no loss of lift. Another example—what happens when an airplane gets into a head wind so strong that it gets "stuck"? Some will always argue that of course it will stall and drop, for an airplane cannot fly unless it has speed. But the experienced pilot knows that it will not stall. Even if you were being blown backward, he says, just simply don't look down at the ground and you won't even notice the wind.

Logic does not seem to mean a thing. Is it easier, you might ask, to fly with the wind than to fly against the wind? Does it take less horse-power, or does it take more? In any group of cadets, some will argue that down-wind flight requires less throttle, because the wind will help to give the airplane the necessary speed. It seems logical, but it is wrong. Some will argue that down-wind flight requires more throttle, because the airplane will have to fly so much faster. It also seems logical, and it is also wrong. And one nationally hooked up quiz program recently explained that for a bird down-wind flight is more laborious than up-wind flight—because the tail wind, blowing it from behind, ruffles up its feathers and thus spoils its streamlining. This is the most ingenious of all, and completely off the target. Actually, any pilot will tell you, wind has no effect on the throttle setting or on power required for flight.

Nothing works out as it ought to work out. In cross-wind flight —does the air press harder against the airplane's windward side than against its leeward side? Common sense says that it does; but a pilot will tell you that it does not. Also in cross-wind flight—should a pilot try to counteract the wind by rudder? Of any batch of cadets, most will say yes, he should. With wind from the left, some will argue that the right rudder must be held to keep the ship from weathercocking, from turning its nose into the wind. Others will argue that left rudder must be held, to keep the ship from drifting away to the right. The experienced pilot says no rudder must be held; he claims that there isn't any tendency for the ship to weathercock, and that any attempt to "counteract drift" by holding rudder is unnecessary, illogical, futile, and even dangerous!

And so it goes, on and on. What about a very heavy, very powerful ship—say a Fortress. Does the wind really affect even such a ship? No, says the student, at least not noticeably. Yes, says the pilot; any ship, regardless of size, weight, or power, is fully subject to the effects of even the slightest wind.

But what are the effects of wind?

IT DOES NOT SEEM RIGHT

In ground school the thing is usually argued out by drawing "vector diagrams" on the blackboard. The arguments are very logical—and very hard work. Some students find the idea of a vector harder to grasp than the wind problem itself. Sometimes an instructor will try to make things clear by comparing the airplane in a wind with a ferry on the river—only to discover that the whole confusion is then simply transferred from air to water, and all the phony-logical student arguments will simply be applied to the ferry instead of the airplane. Often, a student will finally understand the effects and non-effects of wind in the simpler conditions of flight, such as straight cross-wind or straight down-wind flight. But he will lose his grasp again as soon as it comes to more complicated maneuvers, such as turns, and the wind effects thereon. At best, the arguments and diagrams often convince the student's brain but leave his nervous system unconvinced. "I can see what you mean," he says, "but it just doesn't seem right."

But if a man wants to be a pilot, wind effects must seem right to him. He had better chew on the problem until they do. For, if they do not seem right, they will produce tenseness; and a tense pilot is a poor pilot. Also they will prevent him from understanding navigation. Finally, they will lead to mistaken reactions that will spoil his entire flying technique and may in some emergency cause that mistake-piled-on-mistakes which spells the end.

Fortunately the whole thing can be understood in one fell swoop, for once and for all. It is not a task for the cold brain, but rather for the imagination; once you can "see" it you no longer need much logic. It is a problem of point of view; once you can look at the wind from the point of view of the pilot, all the answers become fairly self-evident. The whole thing hinges on three key ideas.

THE FIRST KEY IDEA: AIR IS A SOUP

The first key idea is the idea of air: that air is real stuff, a thick and heavy fluid, quite similar to water. To a pilot, this seems obvious; but to the general public, and hence to many students, cadets, and so on, "air" means hardly more than another word for "empty space"—void. Therefore they cannot understand what wind really is. Wind means to them merely a mysterious "force" (represented perhaps by an arrow on the blackboard) that comes somehow blowing through space and hits the airplane. Or perhaps, absurdly, wind means something that comes whistling through the air! This of course makes no sense whatever; it hardly bears writing down. No one would think it if he stopped to think; yet it is a fact that many a student pilot tries to base his flying on such flimsy notions.

Actually, "wind" is of course simply the fact that the air fluid is in flow. We could see that more clearly if our vocabulary did not confuse us. The mere word *wind* is itself a cause of trouble. We call the stuff *air* when it lies still, but we call it *wind* when it is in motion. That is like calling an automobile an automobile when it stands still, and then calling it, say, a *hickey* when it gets going. It is the same thing, moving or still. Wind is air in motion.

THE SECOND KEY IDEA: MOTION IS RELATIVE

The second key idea has nothing to do exclusively with the air but is a general one that might be called *the relativity of motion;* you have to limber up your ideas of what motion really is. Skipping all fancy business, it will be useful to observe a man who is walking about inside a moving railroad train. This is because he is just like an airplane flying in a wind, as will be shown; but while the airplane in a wind is puzzling, you know or can easily try out just exactly what happens to a man who walks about in a moving train.

That man can walk about in the train just as if the train were standing still. Except for some jolting and bouncing, its headlong speed, of, say, 60 m.p.h. does not affect him. He finds it requires no more effort to walk toward the front of the train than to walk toward the rear; if he did not look at the moving scenery outside, he could not even tell which way the train is moving. If he drew the blinds, he could easily persuade himself that the train is moving in a reverse direction. It is important that you believe this; if you don't, try it and watch. If the track were smooth, he could not even tell whether he is moving at all! Proof of this you have sometimes experienced while in a station, when the train on the adjoining track began to slide out, and you, in momentary confusion, had to make a conscious effort to determine whether it was your train moving, or the other train, or perhaps both of them!

We disregard here the forces that pull on the passengers while a train starts up and accelerates, and again while it brakes; and we disregard also the jolting due to roughness of track. We are talking here only about steady motion. As long as the train maintains a steady speed, the passenger can step from the right-hand to the left-hand side of the coach (walking "cross-train" as it were) without making any allowance for the motion of the coach itself—entirely as if the coach were standing still. Although he is then going sideways at 60 m.p.h. he need not do any special balancing to stay on his feet! In short, in some respects the motion of the train simply does not concern the passenger; steady motion feels to him like no motion at all.

When a pilot flies in a wind, the wind is his railroad coach, but we shall see about that later.

In another respect, the train's motion does very much concern the passenger—the little matter of being carried to Chicago at 60 m.p.h. Regardless of whether he walks, runs, sleeps, dances, or stands on his head within the train, the train carries him right along. The results, if you put them bluntly, are quite paradoxical. For example, he might walk all the way forward to the dining car, have dinner, and walk all the way back to his seat, and he might then say, "Well, here I am, back where I started from." In one sense he would be right, yet in another sense he would be wrong. The place to which he has returned has itself in the meantime moved something like 30 miles. Whether or not he is truly back where he started from depends, in highbrow words, on his "frame of reference," that is, which side of the facts he chooses to regard, and which side he chooses to disregard; whether he judges his position by reference to the train, or by reference to the United States outside.

You could keep on constructing complicated examples of this sort. Suppose the train is sliding slowly out of the station. The fellow's best girl is on the platform seeing him off; he walks toward the rear of the train and temporarily succeeds in standing still relative to his girl. There are many similar examples. Nor is this relativity of motion peculiar to railroad trains; you find the same story again when a ferry crosses a river or when a man swims in a current, and perhaps you used to have fun with it as a kid when you walked down the up-escalator in your local department store, watching yourself walk down and yet stay up. It is all perfectly familiar; the only difference is that in flying it suddenly matters.

THE THIRD KEY IDEA: YOU'RE *in* THE AIR

For here is the third key idea—an airplane that flies in moving air is like a man who walks within a moving train. This, mark well, is not a simile, not a figure of speech, but a precise statement of fact. The two cases are not only pretty nearly alike, but they are just plain alike. The mathematics, physics, and logic of the two cases are identical. Just as the passenger is "contained" by the railroad coach and cut off from the scenery outside, so is the airplane, once in flight, entirely "contained" by the surrounding air and cut off from all connection with the ground. Its propeller propels it, not by acting on the ground, but by acting on the surrounding air. Its wings hold it up, not by acting on the ground, but by acting on the surrounding air. Thus the cubic mile or so of air in which the airplane flies is to it exactly what the railroad coach is to the train passenger. The only difference is that the railroad coach surrounds the passenger visibly, while the air surrounds the airplane invisibly.

Hence, whatever is true of a passenger who walks within a moving train is equally true of an airplane that flies within a mass of air which is itself in motion. The airplane, too, has two motions, both at the same time. It has a motion *through* the air, which corresponds to the walking of the train passenger through the train. It has a motion *with* the air, called *drift*, which corresponds to the travel of the train. Just like the train passenger, the pilot can fix his attention on one sort of motion or on the other, as he chooses; and after he has become at home in the air, he can watch both motions at the same time and not get confused.

MOTION *through* THE AIR

Look first at his motion *through* the air: in a certain sense, this is the all-important motion, since it is what makes the air strike his wing and produce lift; what makes the air strike the tail surfaces and gives him stability and control. And incidentally, it is the motion which registers on the air-speed indicator.

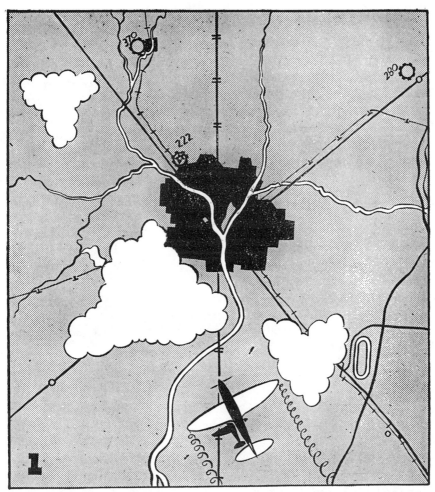

Crabbing: Cross-wind flight, observed from a stationary vantage point high above the country. In this picture the air is partly visible—remember, clouds are nothing but visible air. Wing-tip vortexes (sometimes visible in real life) have been made visible in this picture to show the ship's path through the air. The pilot intends to follow the railroad track that goes straight up the middle of the picture. He points his ship, however, at the T-shaped cloud at the upper left. "Wind" "blows" from left to right. Now look at pictures 2 and 3 turning pages rapidly to get cinematographic effect.

This motion of the airplane *through* the air is in no way affected by the fact that the air itself is in motion—any more than the passenger's walking within the train is affected by travel of the train. The train passenger finds that it is just as easy to walk forward in the train as it is to walk toward the rear or from one side of the train to the other: it requires no different balancing, produces no different sensations. In the same way, the airplane cannot "feel" any difference between down-wind flight and up-wind flight and cross-wind flight. It feels the same; the engine has to pull no harder; the air-speed indicator indicates the same; the lift is the same.

THE AIRPLANE DOES NOT FEEL THE WIND

If the train passenger does not look at the moving scenery outside, he cannot tell which way the train is moving; if the pilot does not look down at the ground, he cannot tell which way the wind is blowing. Like the train passenger, the pilot can even draw the blinds; simply get above a solid overcast, so that you cannot see the ground, and you will not know whether you are flying in a gale or a calm. In short, the wind does not have any effect on air speed, lift, stability, and control: an airplane, once in flight, cannot "feel" the wind.

MOTION *with* THE AIR

Look next at the airplane's moving *with* the air. This is exactly the kind of motion that a train passenger performs by just simply sitting still—or rather the kind of motion that he performs, because of being in a moving train, regardless of what antics he may cut within the train. Regardless of what maneuvers the airplane executes within the mass of air that surrounds it, it helplessly participates in the motion of that air mass; it "drifts" with the wind.

Beware of thinking that the airplane drifts because the wind is blowing against it. It would be like saying that the passenger gets to Chicago because his coach pushes, shoves, kicks, and pummels him to Chicago. It implies that the wind is pressing harder against one side of the airplane than against the other. This is not so; the wind can't blow against the airplane, because the airplane doesn't present itself to be blown at; it yields without any resistance: it moves *with* the air.

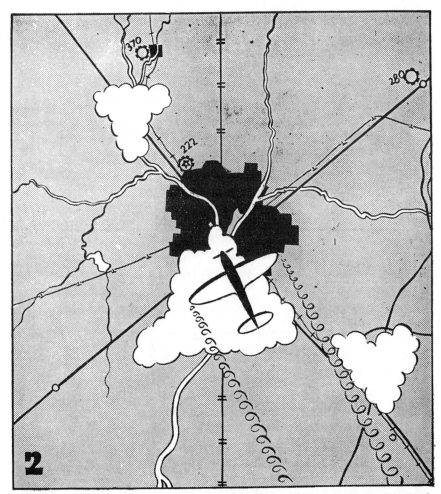

Compound motion: Ship has progressed through the air toward T-shaped cloud. Straight wake behind it shows that it is flying straight. The whole air mass has meanwhile flowed from left to right; the heart-shaped cloud now covers the race track, the big cloud is over town. The ship's motion through the air, shown by its wake, compounded with the air's own motion across the ground, shown by present position of clouds, has moved the ship straight up the railroad track, as intended by the pilot.

But to an observer who could not see the air, the ship would seem to be sidling.

A pure example of this kind of motion, drift, is the travel of the free balloon. Here is one kind of aircraft that has no motion at all *through* the air (other than perhaps climb or descent) but moves only *with* the air. The free balloon has no "air speed," no "relative wind." To its pilots, it always seems to be flying in a complete dead calm. Even though it may be flying in a gale and drifting across the country at 50 m.p.h., there still won't be even the gentlest breeze in its basket. This is because it yields so completely to the motion of the air, and participates in it: the balloon does not push against the air, and neither does the air push against the balloon; hence no forces arise. Flight students sometimes argue that the wind must be exerting forces on a drifting balloon—else why would there be any drift? The argument is drawn from a mistaken understanding of elementary physics, which could be cleared up by rereading pages 1 to 10 of almost any college text on physics. The answer is that wind cannot exert any forces on the drifting balloon, precisely because the balloon is drifting. Or, by expressing the same thing in other words, the balloon drifts precisely because drifting will nullify all wind forces on it.

COMPOUND MOTION

Altogether, then, the path that the airplane takes over the ground is always compounded of those two separate types of motion: its motion *through* the air and its motion *with* the air. The two are entirely dissimilar. Motion *through* the air produces lift, drag, stability and control. Motion *with* the air is the free-balloon sort of motion—it has no further effects on the airplane other than to move it. The two are dissimilar; yet to the eye they are indistinguishable. The eye, which cannot see the air but can judge only by reference to the ground, simply records the compound of the two—the resulting motion of the airplane relative to the ground.

This causes some confusion for the pilot—especially for the beginner. Even though he may understand it in theory, he is still only a ground animal, not an air animal. He still can see only the ground, not the air. His sense of balance, his sense of motion, his nervous system, tend to react simply and naïvely to his observed motion over the ground, whether he wills it so or not. But, since his motion over the ground includes both motion *with* the air and motion *through* the air, his resulting control actions are bound to be wrong. He

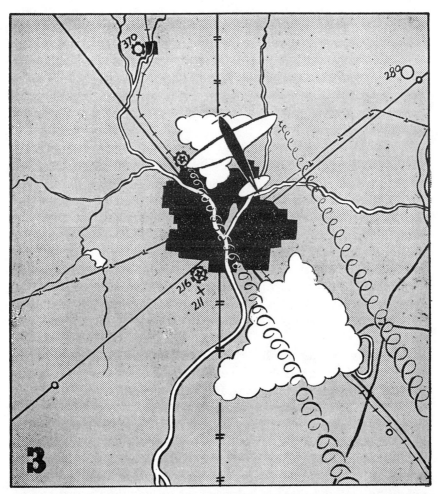

Successful speculation: The ship has now reached the T-shaped cloud toward which it has been headed all along. Meanwhile the whole air mass (of which the clouds are only a visible part) has flowed on farther. At the moment when he has reached the cloud, the cloud is over the railroad track. Thus he is making good the intended ground track. Note that his path through the air has been straight, as shown by the ship's wake. Cross-wind flight requires no rudder, nor any other fidgeting with the controls. The airplane does not feel the wind.

must learn to discriminate between the two and to discriminate continually.

SOME CASES

It may be useful here to discuss the most frequently encountered drift effects—exactly what goes on, how it feels and looks to the pilot, what a man's "natural" reaction would be, and what the pilot's reaction ought to be.

CRABBING

Case number one is straight cross-wind flight. This is the simplest of all wind-drift cases and usually the one where the student first discovers that there is such a thing as drift. Suppose there is 20 m.p.h. wind from the west; suppose you are right above a railroad track that runs true north. Suppose you head your 100 m.p.h. airplane so that its nose points straight north; you fly a quarter of an hour and where will you be? You will be 25 miles north of where you started, owing to your motion *through* the air. But you will also be exactly 5 miles east of the railroad track, owing to your motion *with* the air.

It could not possibly be otherwise. The airplane, being headed north, works its way northward through the air. And, being completely embedded in the air, imprisoned in the air, it must at the same time participate in the air's eastward flow. It could not be otherwise; but the trouble is, it looks wrong.

The ground-accustomed eye naïvely records that the airplane is sliding sideways. To the ground-accustomed nervous system, this motion seems unhealthful. Certainly if a car ever moved in this fashion, it would be time to do something about it. And thus the ground-accustomed sense of balance goes to work and wants to stop that sideways motion by fidgeting with the controls.

Usually, the student has a strong itch in his foot to use left rudder to stop the crabbing to the right. This is not only quite unnecessary; it is also impossible. By using left rudder he will merely yaw the airplane over to the left, making its nose swing over farther and farther; but the airplane will even then fail to go where it is pointed, and the rightward sidling motion will continue. If the student kept on holding rudder in this attempt to "counteract the drift" he would finally turn the airplane completely off its intended heading, would

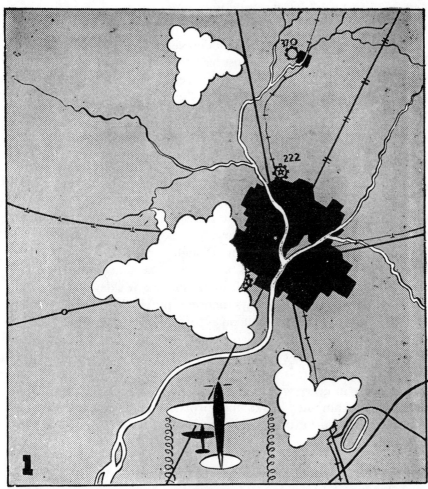

A matter of viewpoint: The following series shows the same story as the preceding one. Same wind, same airplane, same flight, "photographed" at the same three moments. Difference: The observer now moves with the air, as if he were in a free balloon. The first series showed drift as a ground observer would see it (except that it was seen from high up). This one shows drift more nearly as the pilot experiences it. This series may be confusing, but so is drift. It's worth a student pilot's while to dope it out.

perhaps go clear around in circles—and still the eastward sliding would continue!

Because he notices that his rudder veers him off, the student is likely to hold right aileron against his left rudder, flying with his right wing low. Thus the two controls cancel each other, and the only result is that the airplane is forced to fly inefficiently, in a continual slight sideslip; and the eastward sliding of the airplane still continues.

Sometimes a student comes to the opposite conclusion. "The wind is obviously blowing at my ship's left side," he reasons, "blowing it over to the right. From what I know about the purpose of an airplane's tail fin, and judging from my experience in cross-wind taxiing, the wind must now be trying to weathercock me around to the left." And so he holds right rudder against a weathercocking tendency that is entirely imaginary.

Note that this fellow's reaction, though opposite from the previous fellow's, is based on the same fallacy, that is, that air is somehow flowing sideways against this airplane, shoving it sideways. But that is not so; the airplane's sideways motion is purely drift, purely motion *with* the air. It is the free-balloon type of motion, forceless. Since it produces no forces on the airplane, it does not call for any compensating work on the controls. And the airplane's flight through its own element, the air, is absolutely straight, safe, well balanced—the ship is flying in every way as if the day were completely calm. That the motions combined make the airplane slide sideways over the ground may be disturbing to the eye but in no way endangers the airplane.

HOW TO COUNTERACT DRIFT

Still, one might argue, all that may be so; but just the same, that sideways motion of the airplane has got to stop. Hang it all, I wish to follow that railroad track and fly straight north. I don't care about the nature of drift—I just don't want to drift.

The answer is that the sideways motion of the airplane cannot be stopped. Nothing in the world will keep an airplane from sliding eastward as long as it is flying in an westward-flowing air. But you can outfox this effect, and the experienced pilot does so automatically. Upon noticing the eastward drift, he points his airplane slightly west of north. It is a speculation, betting that, when the eastward motion

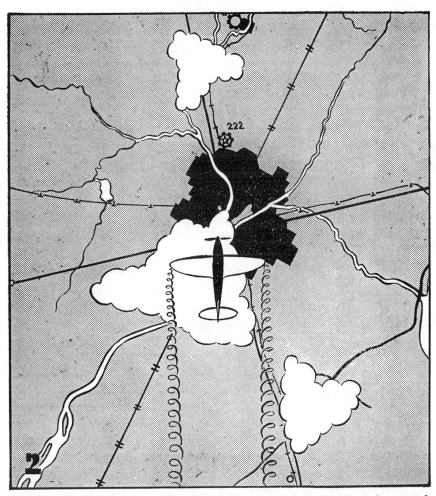

Straight through the air: The pilot flies perhaps without even looking at the ground. Perhaps he has no special ground track in mind which he wants to follow. He just flies. If the clouds were solid under him, he would not even notice that there is a wind, If he does watch the ground, he will notice that the ground does not move as it "should."

with the air adds itself to this northwesterly motion *through* the air, the resulting motion over the ground would be exactly north. The experienced pilot, with his skilled perceptions, makes this speculation with great sureness. And he discovers instantaneously (by watching the apparent motion of the ground) whether he has allowed too much or too little for drift and makes any necessary corrections almost unconsciously. The student pilot is of course less sure of eye. In his case, it becomes more apparent that "allowing for drift" is a speculation; you head the airplane around by guess, and then you wait a while, watch its track over the ground, and see whether it goes where you intend to go. If it does not, you make a new correction and head the airplane slightly differently, until finally you find the heading which, when combined with the drift effect, makes you go where you want to go.

It is important for a student to understand clearly that this turn (by which he heads the airplane around so as to allow for drift) is a normal turn, made in the normal manner, by banking the airplane, using the rudder only to take up the adverse yaw effects of the ailerons. Beware of making these corrections by just kicking rudder.

And it is extremely important for the student to understand that, once the turn is made and the airplane points in some new direction, flight *through* the air is then again perfectly straight; there is then no need for any rudder or any other fussing with the controls. The fellow who thinks he must hold rudder "to counteract drift" does not know what it is all about.

COMING IN DOWN-WIND

Number two of the wind-drift cases is straight down-wind flight. Ordinarily, at altitude, there isn't much of a problem here. You simply make good time, and that's all. But suppose that a pilot, fatigued or absent-minded, misreads the wind sock and deliberately makes his approach to land down-wind, thinking he is landing up-wind. That fellow can easily get confused by wind effects.

The ship is fast, gliding through the air, at, say, 75 m.p.h.; it is at the same time drifting with the air at, say, 20 m.p.h. And since the two motions are both in the same direction, the ship's speed relative to the ground is 95 m.p.h. "Boy," thinks the pilot as he looks down and sees the ground flow by under him, "I am sure going some.

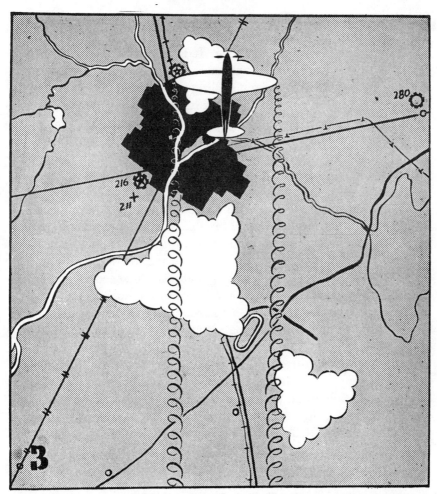

There seems to be a wind: The pilot has flown straight through the air all along. But the town which was originally on his right is now on his left; and has failed to fall behind him as it "should." The pilot decides that he has a quartering head wind which is drifting him to the right and reduces his ground speed.

This is one time when I am in no danger of stalling. In fact I guess I had better slow this thing up a little."

Poor fellow. He is quite wrong. It is true that he has plenty of speed. And it is also true that speed is what keeps an airplane from stalling. But what keeps it from stalling is not speed pure and simple, but speed *through* the air: the air stuff has to keep flowing against the wing. Of the fellow's 95 m.p.h. speed at this moment, only 75 m.p.h. are motion *through* the air (air speed) and lift-producing. The other 20 m.p.h. are motion with the air (drift) and from the standpoint of lift-making they are fake speed; they do not help to keep the airplane up.

It may even be that the pilot understands this quite clearly—in theory. But because he does not realize that he is coming in downwind, his understanding does not help him much; and in any case, he tends to go by what he sees, namely, fast motion of the ground. He ought to be feeling in his controls that his air speed is not at all excessive. The position of his nose relative to the horizon ought to be telling him the same thing; also the sounds of flight ought to warn him. But man, the ground animal, is an eye animal. In the inexperienced pilot, especially, visual impressions tend to override all other perceptions. And so the poor fellow slows himself up. And then a little later he is very much surprised when, with all his speed, he all of a sudden feels stalled and sees himself settling. If he is bright, he then dumps his nose, slams on his throttle, recovers, and climbs away to think things over. If he is not, he hauls back on the stick, and the papers report yet another airplane that was seen approaching for a landing, took a sudden nose dive, and crashed.

WIND EFFECTS ON TURNS

The effect of wind on turns can be best understood through a training maneuver sometimes called the *wind-drift eight*. You are headed straight south and are flying straight and level. At the moment when you pass exactly over some conspicuous object on the ground, for example, a barn, you go into a turn of, say, 30 degrees of bank. You hold this turn for 360 degrees, that is, all the way around, until you are headed south again, holding the bank carefully constant. In still air this would mean that you would fly a circle, and in still air it would therefore bring you back exactly over that barn. In a wind it works out differently. Even though you hold absolutely constant bank and thus describe an exact circle *through* the air, the air in which you fly moves on, and you within it; the path described over the ground is not a circle but looks more like a figure six. At the moment when you are headed south again you are not over the barn but a couple of hundred of feet down-wind of it.

If you can understand this maneuver you understand the whole problem of drift in turns. It is exactly the case of the railroad passenger who leaves his place, goes to the diner, comes back to the old place and yet, judged by the geography outside, is not really back where he started from, but some 30 miles farther on toward Chicago. You come back to the same spot of air that you flew through when you started the maneuver; perhaps you come back so accurately that you feel your own wing wash and slipstream still swirling there from your previous passage through that air spot. But that spot of air has itself meanwhile moved on a couple of hundred feet.

That is the logic of the thing; and of course, the drifting of the ship during the maneuver is not accompanied by any forces, sideways, forward, or backward. It is accompanied, however, by confusing sensations for the pilot; that is why it is a good training maneuver. If the pilot did not watch the ground, he would not notice any wind effects and would not get mixed up. But he cannot help seeing the ground and noticing how now it speeds up, now it slides sideways, now it slows up, and now it slides to the other side. Just how he will tend to react seems to depend on just where his attention is centered. If it is centered on his path over the ground, he notices how this path is being pulled out of shape, and involuntarily he tries to keep it in shape, maybe by shallowing and steepening his bank, maybe—and this would be really bad—by trying to "hold rudder against the drift" and slipping or skidding during the parts of the turn when he has the "wind" from his side. If his attention is centered mostly on flying a nice turn, the skidding and slipping tendencies will be opposite; as he notices the ship's apparent sideways sliding, he will tend to skid as he turns from up-wind to down-wind and to slip while he turns from down-wind to up-wind. In either case, he must learn simply to relax, let the drift effects take place, and make no attempt to resist them. He must learn to fly the ship by attitude and the feel in the seat of his pants mostly and to disregard the strong impressions of sideways and improper motion that the ground sends up to his eyes. A pilot must judge things less by eye and more by his other senses than the ground man.

Some training maneuvers set the pilot the opposite task: to fly a perfect circle over the ground despite wind effects. This is the main idea of most of those figure eight training maneuvers that are so heartily disliked by students. The exact requirements of those maneuvers are subject to change as government inspectors and flight instructors change their ideas. But this is common to all of them: if

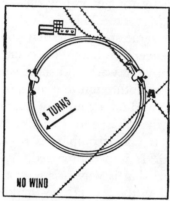

In calm air the pilot who describes a perfect continuous circle *through the air* also describes a circle *over the ground*.

there is a wind, you cannot fly a perfectly circular track over the ground by flying a perfectly circular flight path through the air. If you fly a circle through the air, you get a distorted ground path, as in the so-called "wind-drift eight." If you want a circular ground path, you must distort your air path.

And since the circular ground path is the main idea of most eights, let us state here how to achieve such a ground path. The *exact* air path flown during such a maneuver is an intricate mathematical problem. But for practical purposes it is accurate enough to say that the more nearly the airplane is headed up-wind the shallower must be its bank. The more nearly it is headed down-wind, the steeper must be its bank. The steepest bank will come at the moment when you are flying exactly down-wind. The shallowest bank will come at the moment when you are flying exactly up-wind. Thus if one divided the circle into two halves, an up-wind half and a down-wind half, as shown in the diagram, the bank would be continually steepening during the up-wind half. It would be continually shallowing during the down-wind half.

SOME POPULAR ARGUMENTS

It may now be worth while to answer some questions that are invariably asked by student pilots. "You claim," someone will ask, "that once an airplane is in flight, it does not feel the wind, but flies, balances, and responds to the controls, exactly as if it were flying in a

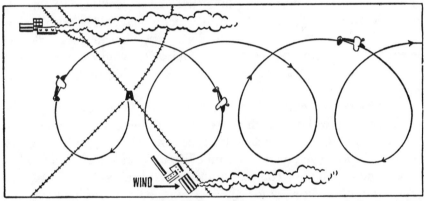

In moving air the pilot who describes exactly the same continuous circle *through the air* describes this pattern *over the ground*.

calm. In that case, why are we required to do spins, chandelles, and so on, up-wind?"

The reason is not that to do them down-wind or cross-wind would be dangerous or difficult; but doing them up-wind keeps students from getting lost while concentrating on the maneuver. If you do, for instance, a series of spins, you will make slow forward progress through the air. If you do them up-wind, then the air itself will at the same time drift backward, and you with it; you will stay fairly well in one position relative to the ground, and when you have finished and look around, your airport will be in sight. If you do the maneuver down-wind, your motion *with* and your motion *through* the air will not cancel each other but will help each other in causing considerable speed over the ground; and, when you look around again, your airport is out of sight.

"You claim that any airplane, even the most powerful, is going to get the full effect of the wind, even the slightest breeze. Flying against a 5 m.p.h. breeze, even a pursuit will lose 5 m.p.h. Why, it seems to me that the pilot would merely have to open his throttle a little more."

Correct. The pilot may be able to maintain his intended ground speed, despite the head wind, by opening the throttle more, retrimming the ship for lower angle of attack, and flying faster. But that would not evade the wind effect. Instead of having a 350 m.p.h. airplane that makes good only 345 m.p.h. over the ground, he would then simply have a 355 m.p.h. airplane that makes good only 350 m.p.h. over the ground. The 5 m.p.h. drift effect would still be with him.

"You claim that in cross-wind flight, any airplane, regardless of weight, size, or horsepower, is fully subject to wind drift—even if a ship weighed 50 tons and cruised at 300 m.p.h.; well, now I just know that can't be right."

You've got something there; but it isn't what you think it is. Yes. a big heavy ship will drift with the wind just as helplessly as a Cub. But speed does make a practical difference (weight, size, or horsepower do not). Suppose you have a 100-mile trip to fly, and a 20 m.p.h. wind is blowing across your course. In a ship that cruises 100 m.p.h., you point the nose directly for the destination, without making any allowance for drift. You reckon that the trip will take one hour. After an hour, you look at your watch and look at the ground, and your destination is nowhere in sight; the ship has drifted 20 miles to the side during the hour—and average visibility in the Eastern United States is only about 7 miles. Now try the same trip in a 300 m.p.h. ship. Point it, too, directly for the destination, without allowance for drift. It, too, will drift in the wind, and in an hour will drift exactly 20 miles. But the trip will take only 20 minutes. When, after 20 minutes, you look at your watch and look at the ground, your destination is off to one side, but still within sight: in those 20 minutes the ship has drifted only about 7 miles. Thus while the fast ship drifts just as helplessly as the slow ship, the drift is proportionately less important to the fast ship; the drift *angle* is smaller. This is one reason why a fast ship is much easier to navigate.

"You claim that a man can walk within a moving train just as if the train were standing still. Well, have you ever tried? Don't you get bounced around like other people?"

This is an important argument. The train in moving rapidly along its tracks hits all sorts of roughness and throws its passengers around. In the same way the air in sliding rapidly across the country hits all sorts of roughness and throws the airplane around; on a windy day the air is almost always rough. But it is still true that you can't tell from a train's bouncing which way it is going. And the same thing is true of the air. On a windy day, the air may be rough, but it still makes no difference whether you fly up-wind, cross-wind, or down-wind. At least that is what aeronautical science teaches now. Some pilots claim that the wind roughness has different effects on the airplane, depending on whether you are flying up-wind or down-wind. This is a moot question; little is known about the internal structure of a gusty wind. Whatever may be the truth, keep in mind that it applies only to gust effects, not to the wind itself. What has been explained in this chapter is the basic facts. People who talk about gust effects on airplanes are talking about a fine point, and all of them believe the basic facts explained here are true. In this connection the reader should also be warned that, near the ground, wind velocity varies rapidly with altitude. An airplane that climbs or descends or banks near the ground is sometimes subject to some special effects that are similar to what happens when you board or leave a moving bus.

"I still insist that an airplane would not drift unless the wind were shoving it." "Yes, and what about cross-wind take-offs and landings?" An object moves, not because a force is acting on it, but because a force has once acted on it. It will then keep moving until another force stops it or gives it some different motion. The time when the air shoves the airplane is right at the take-off. On a cross-wind take-off, right after you break ground, you can for a couple of seconds actually feel the sideways shoving of the air, shouldering you over, accompanied by a weathercocking tendency. But it lasts only a couple of seconds; by that time the airplane has yielded to this force, and from then on it is true that an airplane cannot feel the wind.

Chapter 7

WHAT THE AIRPLANE *WANTS* TO DO

"**F**ISH got to swim, birds got to fly"—and the airplane, too, can't help itself; it has to fly, for it is made that way. Take your hands off a good airplane's stick, and it will do a good job of flying all by itself. Take your feet off its rudder as well, and many airplanes will even then do a fair job. The airplane has a built-in will of its own, and generally speaking it wants to do whatever is necessary to maintain healthy flight.

This built-in will of an airplane is technically called its *stability*. An airplane is stable if it wants to do the right thing, unstable if it wants to do the wrong thing. That the modern airplane is stable (in most respects) rather than unstable, that it wants to do the right thing rather than the wrong thing, is one of the most important facts behind the whole art of flying. Not that we particularly need to fly hands off. But a stable airplane's ability to fly hands off is a sign that, even when the pilot controls the airplane, the airplane helps him—instead of fighting him as it well might, and as some of the early airplanes did.

If the art of flying consists in the very first place of doing nothing—rather than of the continual frantic balancing act which it would be otherwise and which many laymen imagine it to be, it is because of this tendency of the airplane to do the right thing anyway. The airplane corrects for small aberrations from healthy flight all the time, nips them in the bud before they can even require action from the pilot—in fact, before they even become noticeable; no, more than that: it does not even let them get started. Furthermore, whether he realizes it or not, the pilot is always being guided by his airplane's feel. Because the ship wants to do the right thing and resists doing the wrong thing, he can always sense whether he is doing right or wrong largely by the feel of the controls. Whatever the piloting tricks are by which we impose our will on the ship—banking it or yawing it, slowing it up or diving it, mushing, climbing, sideslipping, stalling, spinning—we are always using them against the background of the airplane's own will. It is the airplane's own will, conflicting with ours, that puts up the resistance we feel as control pressures.

For example, when you slow the airplane up, it becomes nose-heavy and makes the stick pull forward against your hand; and it is largely by this pulling that you know you have slowed it up! Anyone who has ever flown an unstable airplane knows how much we really need that sort of guidance, how badly we miss it if it isn't there or can't be trusted. An airplane, for example, that becomes slightly tail-heavy when slowed up (thus wanting to do the wrong thing, tending to stall itself), seems quite treacherous and is very much harder to fly well. "You have to watch it all the time." During an approach, for instance, when the pilot's attention must be on the ground, on the glide path, on traffic rather than on the airplane itself, it is easy to stall such an airplane inadvertently simply because the stick does not remind the pilot to keep speed, does not tug forward against his hand.

But just because the stability of the airplane is so basic a thing, so ever-present, many a student pilot never bothers to find out exactly *what* it is the airplane wants to do. Offhand, he will probably say that the airplane wants to keep its nose on the horizon, that is, that it wants to keep from climbing or diving; that it wants to keep its wings level; and that it wants to keep flying straight ahead.

On each of these three counts, he is wrong.

THE MAIN FACTS OF STABILITY

That a good airplane, left to itself, will fly straight and level and hold its wings level is simply not true; although the statement is found in many books on flying. In the first place it simply is not a true description of what the airplane will do if left to its own devices. The airplane will *not* keep its nose level except under certain conditions, the most important of which are that there must be no bank and that the engine must continue to run at constant power. Nor will the airplane keep its wings level—except when given positive help from the pilot; if the ship is to keep its wings level the pilot must keep the ship straight by positive action on the rudder. And very few airplanes will very long fly straight ahead when the pilot's feet as well as his hands are off the control. Even if "torque" is absent, or is properly compensated for, most airplanes will soon go into a turn, and the turn will become a downward spiral—often a vicious spiral dive.

In the second place, this straight-and-level idea of stability is misleading also in a deeper sense; it tries to describe the "will" of the airplane in the wrong terms altogether. The airplane is not concerned over its own *attitude;* it does not particularly "care" about the earth or whether its nose is pointing straight down or straight up or at the horizon; or whether its wings are level or banked. The airplane is concerned with something else altogether—the relative wind. It wants to keep itself properly lined up, not with the earth or the horizon, but with the relative wind.

THE SPEED-KEEPING TENDENCY

This probably sounds confusing. The best way to make it clear is to wade right in and take as an example the most important aspect of the airplane's stability: *longitudinal* stability, or what the airplane wants to do in the nose-up, nose-down sense.

It is not true, then, that the airplane wants to keep its nose on the horizon or that it wants to maintain level flight. Neither would the opposite statement be true, for example, that the airplane wants to climb or to dive. What the airplane wants to do cannot be described in terms of attitude at all. *The airplane wants to fly at a certain speed;* it wants to keep the relative wind coming at itself steadily.

At least, that is a simple explanation that covers 95 per cent of the answer. Because it is much easier to grasp than the 100 per cent truth, much easier to express in terms familiar to the pilot, we shall discuss it first; afterward we shall give a more sophisticated explanation that gives a more precise answer but is less easy to grasp.

A stable airplane, then, wants to fly at a certain *speed*. It will point its nose up, level, or down as necessary to maintain that certain speed. If slowed up, it will nose down until it has regained that speed. If dived to excess speed and then given its head, it will nose up as much as necessary to get rid of the excess speed and regain the original speed. If the pilot throttles back his engine, the airplane will nose down enough to keep that speed. If in cruising flight the pilot gives it wide-open throttle, it will not speed up but will nose up and climb just steeply enough to keep from speeding up. An airplane of perfect stability, once "trimmed" for a certain speed, will always fly at that speed, regardless of power. With power full on, it will climb at that speed. With just the right amount of power on, it will fly level at that speed; with power off, it will descend at that speed. Thus an airplane may sometimes carry its nose very high and at other times carry its nose well down and yet be stable.

"Well," a student might say, "that doesn't sound to me like any kind of stability at all." It is true that "stability" is defined in physics as the tendency of a body, upon disturbance, to return to its original state. The classic example is the pendulum, a weight hanging on a string which upon disturbance will always finally return to its original position—hanging at rest straight down from its point of suspension. But the "state" to which stability tends to return a thing need not be a position; it may also be, for example, a temperature or a direction or some other condition. In the case of a stable airplane, it is a speed; the airplane always wants to return to its original speed.

All this will make more sense if you imagine for a moment that some fool had built himself an airplane that was stable in the sense of wanting to hold a constant position or *attitude*—wanting to keep its nose level regardless of the circumstances of flight. Consider how such an airplane would behave if its engine suddenly quit. It would act exactly as the very green student tends to act in the same situation.

To such a student speed and Relative Wind and Angle of Attack and lift mean as yet nothing. But attitude means a lot; he wants to see his airplane in a nice, safe-looking, conservative level attitude—never too steeply banked, never nosed up or nosed down. And so this green student, if his engine quits, refuses to let his airplane point its nose down; and, if the instructor does not prevent it, he will presently lose speed and stall and spin and crash.

In the same way, an airplane that foolishly tended to hold a constant *attitude*, would tend to stall itself if its engine should quit. But an airplane that wants to hold a constant *speed* will want to nose down the moment its engine quits. It will thus never stall by its own will; and, if the pilot, in a foolish anxiety to keep its nose level, should force it into a stall, it will warn him and try to resist him by getting noseheavy and tugging forward on the stick.

How does the designer endow the airplane with this speed-keeping tendency? Here again the full story is exceedingly intricate, and it may be better to start with a simplified story, sacrificing some of the truth. Such a simple description would not work if you wanted to build an airplane. But it is true enough so you can fly by it; it allows you to guess what the airplane will do in such-and-such a situation. It is the sort of truth that philosophers call pragmatic truth; a theory is true if you can act as if it were true and not come to grief.

Here, then, is a simple working explanation of longitudinal stability. The front end of the airplane is arranged so as to render the ship noseheavy. You might say that this is why the engine of an air-

The horizontal stabilizer can be understood also as a device to keep the airplane at constant speed. *Top:* Front end of airplane is arranged to be noseheavy thus tending to nose the airplane down. The stabilizer is set to develop a down force, thus tending to nose the airplane up. At some certain speed, the nosing-up tendency due to stabilizer just equals the nosing-down tendency due to heavy nose. That speed is the speed at which airplane "wants" to fly, and to which it will return after disturbance. *Left:* If excess speed develops, the tail force increases, while front-end noseheaviness remains about the same. The airplane noses up, thus reducing its own speed. *Right:* If the speed becomes too slow, the tail force fades out, thus allowing the heavy nose to point airplane down. Airplane then picks up new speed.

plane is mounted so far forward; the airplane's center of gravity must be so far forward with respect to the wings that the airplane as far as its front end is concerned will normally have a tendency to nose down.

THE LEAST UNDERSTOOD PART OF THE AIRPLANE

The tail of the airplane is arranged to resist this diving tendency. The horizontal tail fin is probably the least understood part of the airplane. Looking at an airplane pass overhead, the layman would say, "Yes, of course. In front, they have a pair of large wings to hold up the thick end of the airplane; and in back, they have a pair of little wings to hold up the thin end of the airplane. That makes sense." But it isn't true. The purpose of the horizontal tail fin is not to hold the tail up, but to hold it down; it is a sort of wing, but a wing set at a negative Angle of Attack so that the air flow on it produces in a normal flight a downward force.

In many airplanes, this is not immediately apparent to the beholder. The wings and the tail will seem to be set at about the same angle of incidence. The answer is that the tail operates in the downwash of air that flows off the wings. Remember, the wings make lift by pushing the air down. The air flows down on the tail and pushes it down more than is apparent to the casual observer.

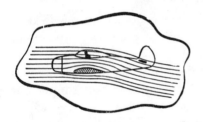

In most airplanes, the stabilizer *appears* to be set for positive Angle of Attack, so as to produce an *up* force on the tail in cruising flight. But actually the stabilizer works in the downwash from the wings. Despite appearances, its Angle of Attack may thus be zero, as in this picture. In most airplanes it is actually negative, producing a *down* force on the tail.

Thus the weight of the airplane tends to nose it down; the horizontal stabilizer tends to nose it up. With the ship properly trimmed for a certain speed (say, cruising), the two forces exactly balance each other. For that is exactly what we do when we trim ship for a certain speed: we adjust the angle of horizontal tail surfaces so that at that particular speed, the downward air force on the tail will exactly balance the downward pull of gravity on the nose.

SEESAW

Now suppose that the engine quits or is throttled back, and the ship slows up. At the slower speed, the nosing-down tendency due to the weight of the nose remains (for all practical purposes) just as powerful as before. But the nosing-up tendency due to the down force on the tail fades out; for this down force is a dynamic reaction of air and hence depends, just as the lift of the wings does, very much on air speed; if the air speed slackens only a little, the down force fades a lot.

The tail then no longer balances the airplane; the nose takes charge and points the airplane down. Once pointed down, the airplane picks up speed; and, as its speed increases, the down force on the tail builds up again until, at the original speed, balance is reestablished; but note that balance is restored in downhill flight.

Suppose now that the pilot opens the throttle wide, and the airplane speeds up; the down force on the tail, being an air force, grows very rapidly as the air speed grows, while the downforce on the nose, being a matter of weight, remains substantially the same. The ship noses up; and, as it noses up and starts flying uphill, its speed slackens. In this fashion, the airplane of perfect stability always seeks *that* attitude and *that* flight path which will allow it to fly at the proper speed.

A MORE SOPHISTICATED VIEW

That the longitudinal stability of an airplane is its tendency to keep its speed constant is not (as some readers might suspect) a private pet theory of the writer's. It will sound strange to pilots, but it is engineering fact, agreed upon by all authorities. It will sound strange to engineers, too; but that is because the engineering fact simply has been translated here from the language of the engineers—which remains meaningless to most pilots—into expressions that may have more meaning for pilots.

In the process of translating, however, one serious falsification has been made. The longitudinal stability of the airplane has been described here in terms of *speed*, but the engineers describe it in terms of *Angle of Attack*. Actually, what the airplane wants to keep constant is not its own speed, but its own Angle of Attack. This makes the whole matter harder to understand, however, and 95 per cent of the time, it makes no difference, as will be shown. Thus the reader may want to skip the following four pages on a first reading of this book.

For the sake of accuracy, however, it seems best to restate the whole matter of longitudinal stability in the more sophisticated terms of Angle of Attack.

In straight flight, including straight climbs and straight glides, the two propositions, one in terms of speed, and one in terms of Angle of Attack, amount to the same thing; if the airplane is flying at a certain speed, it also has a certain Angle of Attack; if it is holding a certain

Angle of Attack, it has a certain speed. It is in *curving* flight that the difference becomes important. In curving flight, the airplane loads itself down, so to speak, with centrifugal force; it then needs additional lift in order to sustain the added "weight." Therefore, if it is to continue to fly at the *same* speed, it needs *more* Angle of Attack; or if it is to continue to fly at the *same* Angle of Attack, it needs *more* speed. Thus, if our understanding of stability is to cover also *curving* flight, it becomes important to understand that actually a stable airplane keeps its Angle of Attack constant and will allow its speed to change if that is necessary to keep its Angle of Attack constant. When an airplane is in a turn and you then release the stick, it therefore will *not* keep a constant speed. In order to keep its Angle of Attack constant despite the "weight" of centrifugal force caused by the turn, it will drop its nose and pick up speed. If the turn is a tight one, the airplane will go into quite a steep dive and go to very high speed, but all the time its Angle of Attack remains the same; in fact it is *because* the Angle of Attack "wants" to remain constant that the airplane dives!

This tendency of an airplane in a turn to drop its nose and to pick up speed in a turn is familiar to every pilot; for in every turn we have to combat it by putting back pressure on the stick—often quite hefty back pressure.

MORE ABOUT THE TAIL

What is the mechanical arrangement that fixes the Angle of Attack at which an airplane will fly and that keeps it so constant? Just as we have redescribed longitudinal stability in more sophisticated terms, we must now reexplain the function of horizontal tail fin of an airplane. The following explanation will seem to be in conflict to the one previously given, but the conflict will finally be explained away.

Previously it was said that the front end of the airplane is always noseheavy and that the tail fin exerts a downward force in normal flight. It was shown how this downward force, growing and fading as the speed grows and fades, acts as a sort of governor of speed. Now, setting that idea aside for the moment, it may be best to think of the front end of the airplane as neutral or indifferent and to think of the horizontal tail surfaces simply as a sort of wind vane.

Consider how it works when the airplane, properly trimmed, is flying level at cruising speed and cruising Angle of Attack. The relative wind blows at the airplane from straight in front. The tail then carries itself in that position where the horizontal tail surfaces will be exactly lined up with the wind, so that the wind will create neither an up force nor a down force on it. In fact, the airplane flies at that particular Angle of Attack because the tail wants to ride in that particular posi-

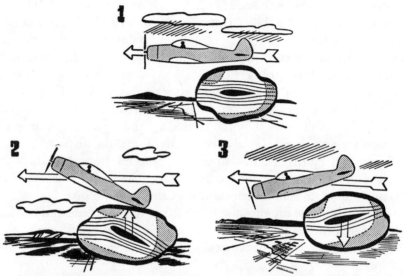

The horizontal stabilizer fixes the Angle of Attack at which the airplane will fly. A simplified picture, disregarding downwash and many other complications. 1. An airplane at normal cruising Angle of Attack. 2. For some reason, the airplane has assumed high Angle of Attack. The Relative Wind now hits the underside of stabilizer and blows the tail up. The airplane returns to the original Angle of Attack. 3. For some reason, the airplane has assumed negative Angle of Attack. The Relative Wind now hits top side of stabilizer and blows the tail down. The airplane returns to its original Angle of Attack. *Note:* in the simple case pictured, return to original Angle of Attack means also return to original attitude. This is not necessarily always so. If power were suddenly reduced, for example, the original Angle of Attack could be maintained only by nosing down. Never confuse Angle of Attack and Attitude!

tion; and the tail wants to ride in that particular position because of the angle at which the tail fin is set on it.

Assume now that the pilot, by turning his stabilizer crank or moving his trim-tab control, sets his horizontal tail surfaces to a slightly different angle. Suppose he changes their setting so that their leading edge is a little lower and their trailing edge a little higher. In the more old-fashioned ships, this is done by actually changing the angle of incidence at which the horizontal tail fin sits relative to the fuselage tail—there being a worm gear by which this angle can be changed. In more modern ships, the horizontal tail fin is usually built rigidly at a fixed angle as part of the fixed structure of the ship; and "trimming" is done by changing to the setting of the elevator trim tab, thus making the elevator ride higher or lower. For the present discussion these technicalities make no difference; the elevator and the horizontal tail fin act as one tail surface. What happens, then, when the pilot changes the angle of that tail surface?

The tail then meets the wind of flight at a slight Angle of Attack and develops a force. If it is trimmed as described above, the angle is negative and the force is downward. This means that the tail will swing down and keep going down until this downward force disappears. The down force will disappear only when the whole airplane finally rides at a higher Angle of Attack. And the airplane will continue to fly at the higher Angle of Attack until the setting of the horizontal tail surfaces is changed again.

Assume now that for some reason the airplane were suddenly forced to a much higher Angle of Attack. For example, suppose that it had gone into a turn, and centrifugal force ("g load") were loading it down, squashing it down, causing it to mush (for "mushing," remember, means nothing but flight at high Angle of Attack). This would mean that the Relative Wind would blow against the wings from slightly underneath and that it would blow no longer against the nose of the fuselage, but more upward against its belly. But it would also mean that the Relative Wind would then strike the underside of the tail surfaces and would blow them upward; the tail would swing up; the nose go down; the airplane's Angle of Attack (its angle to the Relative Wind) would go back to what it was before the disturbance occurred; at the same time, its *attitude* would be more nose-down; and its flight path would be downward, and its speed would pick up. Thus the airplane simply would not balance at the higher Angle of Attack. It would balance only at the Angle of Attack for which its horizontal tail is "trimmed."

Consider still another case. The airplane is flying level at a certain speed and a certain Angle of Attack. The wind then blows at it from straight in front, and the tail is riding a position where the horizontal tail surfaces are lined up with the Relative Wind, getting neither an upward nor a downward force. Suppose then, that the pilot closes the throttle; as the speed begins to fade, some of the lift fades, the airplane begins to "mush" downward, that is, it sinks. But when it is sinking the relative wind no longer blows at it straight back from in front; it now blows at it from slightly underneath, blowing upward against the airplane's belly and against the underside of its wings. But if the Relative Wind hits the airplane's belly and hits the underside of its wings, then it also hits the underside of the horizontal tail surfaces. This blows the tail up, nosing the airplane down until the airplane is again presenting itself to the Relative Wind at the original Angle of Attack. At the same time, the airplane starts picking up speed—so that it will then simply remain, in gliding or diving flight, at the same Angle of Attack and the same speed which it had originally in level flight with power on.

In short, the horizontal tail surfaces act really just like a wind vane —except that this weather vane is sensitive, not to whether the wind blows at it from the east or the west, but to whether the wind blows at it from a little more underneath or a little more in front or above. In any case, it always lines itself up with the relative wind, and thus noses the airplane up or down as necessary to keep the wings meeting the air at the same Angle of Attack.

This now leaves the reader with two different ideas of the airplane's horizontal tail: one, that there is always a down force on the tail; two, that the tail always seeks to ride in that position where there is *no* force on it. The truth is a combination of both. It is true that in most airplanes, in most conditions of flight, the tail surface exerts a downward force as explained earlier. But it is also true that the weathervaning action of the tail, of which a simplified description has just been given, goes on all the time. In truth, the two kinds of action are superimposed upon each other and act concurrently, and both act to keep the airplane *going*. That, in the last analysis, is the essential function of longitudinal stability: an airplane can't fly unless it has speed—unless the air is flowing against its wings all the time. And it is the horizontal tail fin's job to keep that all-important air flow coming —evenly.

COMPLICATIONS, EXCEPTIONS, AND PARTIAL RECANTATION

What has been said so far concerning the airplane's stability in the nose-up, nose-down sense has been an ideal picture, not a true picture. Very few airplanes, if any, actually behave quite so well: very few will actually fly, with the stick released, in straight flight, at constant speed regardless of power. The one ship that comes nearest the ideal is Fred E. Weick's *Ercoupe*, which is in so many other respects, too, a remarkable airplane. It may seem absurd to set up a standard of stability which practically no airplane can live up to; at first glance, it will seem so unrealistic as to be useless. But there simply isn't any other standard of stability. If the idea of stability (longitudinal stability) has any meaning, then it is this speed-keeping, (Angle of Attack keeping) tendency. There is nothing else, nothing but speed or Angle of Attack, you would even *want* your airplane to be stable about. If almost no airplane actually behaves as has been described, that simply means that we are not yet able to build completely stable airplanes. And even so, the idea of stability that has been presented is useful to the pilot. For as he gets into a new ship, this idea gives him a starting point from which to measure the new ship's characteristics: to what extent is the ship stable, and in what particulars does it fall short of perfect stability? If you can answer those questions about an airplane, you really know that airplane.

It may be useful now to look a little more closely at those various imperfections of an airplane's stability—especially at those which are found in almost all airplanes and are the rule rather than the exception.

HUNTING

In actual practice most airplanes have a tendency to *hunt*. This means that with the controls released, the ship will not fly at constant speed but will oscillate up and down, now dropping its nose, building up speed, diving for perhaps 10, 20, 30 seconds. Then, with its essential stability and better self asserting itself, it will catch itself, raise its nose, try to get rid of its excess speed, and go into a climb. In this climb, a bit too much speed is lost, and the ship noses down again, to repeat the process. If the pilot stays off the controls, this may result in a continual roller-coaster-like motion which is an extremely complicated interplay of changing speeds, flight paths, directions of the Relative Wind, Angles of Attack, coupled with changing drags, changing propeller efficiencies and horsepower outputs, changing lifts, and slight downward and upward *g* loads as the airplane alternately pulls itself up from its shallow dives and pushes itself over out of its climbs.

All the time, however, such an airplane is really "trying" to do the right thing. It is trying to get back to its proper speed and Angle of Attack. Its trouble is merely that it "overcontrols" itself and doesn't quite succeed in steadying itself down.

Because such a ship is always trying to do the right thing, it can't be called simply unstable. If it were unstable, it would go into a dive and stay in the dive, or else it would go into a climb, steepen its climb until it had stalled itself, and then in an extreme case it would hold itself in the stall and lock itself in the ensuing spin. Because it is actually always trying to do the right thing, such an airplane is called statically stable—a confusing word, instead of which "primarily stable" would be more useful for pilot instruction purposes.

Some airplanes never succeed in steadying themselves. Their oscillations become greater and greater, and if you leave such an airplane to its roller coasting long enough, it will gradually become vicious, diving to dangerous excess speeds, climbing up into a stall, and then diving again viciously. Such an airplane, "statically" stable, is at the same time called "dynamically" unstable—another confusing word for which "secondarily" might be a handier substitute. Most airplanes, however, are "dynamically" stable; their oscillations will become weaker and they will finally steady down, provided the air is smooth and no new disturbances occur.

All this is usually stressed quite heavily in Theory of Flight courses for pilots. But it is not really too important to the pilot. For, as has been pointed out, stability is important to the pilot not because he actually wants to fly hands off; it is important to him because of the way in which it affects the feel and behavior of the ship even while he controls it. And, while the pilot is at the control, these oscillations will not develop. That's why the whole subject of "dynamic" versus "static" stability is not worth the headache it causes to the student pilot.

THE EFFECT OF POWER

Another imperfection of stability is of the utmost importance to the pilot; yet this one is rarely even discussed. It is the effect of throttle setting upon the airplane's speed, Angle of Attack, "trim"—in short, on its stability in the nose-up, nose-down sense.

It has been explained that in an airplane of perfect stability power does not affect speed. Such an airplane, once trimmed for flight at, say, 100 m.p.h. will fly at 100 m.p.h. regardless of throttle setting. With throttle closed it will descend in a 100 m.p.h. glide hands off. With the throttle partially open it will fly level at 100 m.p.h. hands off; with the throttle wide open, it will climb at a 100 m.p.h. air speed. The usual airplane, unfortunately, is likely to behave quite differently. Trimmed for 100 m.p.h. at cruising power it will, with power off, go into a glide at perhaps 130 m.p.h.—a glide so steep and fast that it must be called a dive. With wide-open power, the same airplane will go into a climb at perhaps 70 m.p.h.

This introduces a peculiar little complication into the art of flying. Formulating it bluntly, let us say: an airplane wants to speed up if you close its throttle, and it wants to slow down if you open its throttle!

The reasons for this are manifold. Propeller blast is one of them. The tail surfaces unfortunately ride in most airplanes in a place where the propeller blast hits them. Thus they get an extra wind in power-on flight, causing an extra down force on the tail. With power off, that wind and extra force disappear. "Thrust-line location" also influences an airplane's reaction to the throttle: where the propeller is mounted in relation to the rest of the airplane—especially in relation to the wings. Consider a high-wing monoplane with an engine mounted low on the nose, to provide better vision for the pilot; obviously the forward pull of the propeller, coupled with the backward pull of the wings' drag, will introduce a nosing-up tendency when the power is on which will disappear when the power is off. Then there is the fact that the tail surfaces ride in the downwash that comes from the wings. With power on, the propeller blast hits certain parts of the wings. Those parts then work in an extra sharp Relative Wind, and send back an extra sharp downwash to the tail, thus producing an extra down force on the tail. And there are other reasons. The whole effect is still baffling the engineers. It becomes the more pronounced the heavier and more powerful the airplane is, and it makes some of the big ones actually unstable in certain conditions of power-on flight! It is thus one of the main differences between trainers and small private airplanes and the big fellows. The pilot need perhaps not understand this effect; but he certainly should realize that it exists.

This effect is important to the pilot because it means that an airplane may under some conditions want to stall itself after all! Suppose, for example, that you have an airplane that cruises at 100 m.p.h. and stalls at 45 m.p.h. You are making an approach, and you have trimmed it for a power-off glide at a normal glide of 75 m.p.h.—so that you need little or no back pressure on the stick to keep that speed constant. Suddenly you notice another airplane taxiing into your path —or a ditch. You open the throttle wide to climb away and go around again. The ship (which "should" now climb at an air speed of 75 m.p.h.) will now actually "want" to fly at 45 m.p.h., that is, it wants to climb like mad right up into a power stall; and until you have retrimmed your stabilizer or reset your trim tab, only very strong forward pressure on the stick will keep the nose down to a safe climbing attitude and the speed up at a safe figure.

The same effect, working less spectacularly and therefore more dangerously, can get you into trouble during an approach—especially in a fairly heavy ship. In such a ship you will prefer, during the glide, to adjust your stabilizer to trim tab so that most of the force goes out of the stick, that is, that the ship will trim in the glide with but little back pressure from your hand. All right then: if during the approach you begin to feed the ship a little power, because you find yourself getting too low, the power will indeed stretch your glide; but it will also slow you up and get you to dangerously high Angle of Attack, close to the stall; it will, that is, if you just keep holding the same amount of back pressure on the stick that you held before you applied power.

To sum up then: Every airplane has a built-in tendency to keep its own Angle of Attack constant, and hence (except in curving flight) to keep its speed constant—regardless of how the amount of power delivered by its engine may change. This tendency is basic, and an essential attribute of the airplane. A pilot simply doesn't understand the airplane unless he understands this speed-keeping tendency. In almost all airplanes, this basic tendency is overlaid by various contrary tendencies, such as "hunting," or the effects of power that have been described in the last few paragraphs. These tendencies are nonessential, unwanted, and purely accidental. It is true that most airplanes cannot be flown safely unless the pilot is aware of these contrary tendencies, but this should not obscure in the student pilot's mind the basic fact that the airplane wants to keep its speed constant.

Lateral, or banking, *stability* is the simplest of the airplane's various self-flying tendencies to understand. Here again, one merely has to keep in mind that the airplane is not concerned with its own attitude relative to the ground. The airplane is concerned only with the manner in which it cuts through the air. It is *not* true, as some student pilots believe, that the stable airplane wants to keep its wings level; often lateral stability will make it bank, and under some conditions a perfectly stable airplane would just as lief bank up to vertical and even past vertical! What the airplane does want to do is to stop slips and skids by rolling against them: the stable airplane wants to hold its wings in such an attitude that it will not slice through the air sideways.

As long as flight is *straight*, this means that the ship will hold its wings *level* and, if disturbed, will bring them back to level; for if it carried one wing low it would start sideslipping to that side. In *turning* flight, it means that the airplane wants to bank "just right"—so as neither to slip nor to skid. If slip or skid does for some reason develop, it means that the airplane will promptly bank or unbank to whatever degree is necessary to stop the skid or slip. For instance, if the pilot simply kicks right rudder, the airplane will swing its nose to the right but keep on flying in the old direction—thus slicing through the air left side first; but its lateral stability will then bank it to the right. Or, suppose that a gust puts the airplane's right wing down and the left wing up; the result of that wing-low attitude will be that the airplane slips off sideways to the right. But as soon as it does so, its lateral stability goes into action and does the thing that is necessary to stop that sideslip; it raises the right wing again.

In short: lateral stability is *not* the tendency of an airplane to keep its wings level. The airplane has no such tendency. Lateral stability is the tendency of an airplane to bank or unbank its wings so as to avoid sideslipping.

This tendency of an airplane to refuse to sideslip is due mostly to its "dihedral"—the V-like uptilt of the wings as viewed from in front.

How dihedral functions in stopping a slip can be shown in a picture more clearly than in words.

Consider the picture on this page; if the ship shown there should be moving not where it is pointed but exactly toward the observer, then it would be sideslipping; and, to stop that sideslip, it would then have

Dihedral is the V-like uptilt of the wings shown here. It makes the airplane bank or unbank as necessary to keep from slithering sideways through the air. *Insert* shows such sideways motion: dihedral stops it by raising the airplane's right wing. *Main picture* shows how dihedral works. If this airplane is moving toward your eye (instead of moving where its nose points), then its Relative Wind streams at it from your eye (instead of streaming at it from in front). Observe how its right wing presents itself to your eye at a high "Angle of Attack" (bottom surface is visible) while its left wing presents itself at a negative "Angle of Attack" (top surface is visible). The wings present themselves to the Relative Wind in precisely the same fashion. Thus the right wing develops more lift than the left wing: the ship banks away from you, and this stops its sideways slithering motion toward you.

to roll and bank away from the observer. And that is exactly what its dihedral will make it do. You can see how: the flight path is toward the observer; the Relative Wind therefore blows at the airplane from the observer. The shape that the ship presents to the observer's eye is also the shape that it presents to the Relative Wind. You can see how one wing presents itself at a much higher Angle of Attack than the other;

it therefore will develop more lift; the airplane will therefore roll itself away from the observer; and the sideslip will stop.

THE WEATHERCOCKING TENDENCY

The airplane wants to do the right thing, rather than the wrong thing, in still another respect: it will swing its nose right or left as necessary to maintain healthy flight. This particular self-flying tendency is called *directional stability*. It is a little harder to understand, but only because it is so completely misnamed. Strictly interpreted, the expression "directional stability" can mean only one thing—that an airplane, once headed west, will keep flying west by its own will; and, if some disturbance turns it in some other direction, that it will turn back and resume westerly flight. But no airplane will do such a trick; not unless it is equipped with some gyroscopic course-keeping gadget such as is incorporated in the Automatic Pilot. If you interpret the expression "directional stability" a little more loosely, you might take it to mean that an airplane, once headed west, will keep flying west by its own will; and, if some disturbance turns it, say, south, that it would then straighten out as soon as the disturbance had passed, and would from then fly south. But unfortunately probably no airplane will do even that trick—although it is claimed for some airplanes. The ordinary airplane, left to its own devices, will *not* fly straight but will start turning, and, once turning, it will not straighten out but will keep circling and will probably even go into a spiral dive—as we shall see.

What this so-called "directional stability" actually does can best be expressed by the term *weathercocking*. It is the tendency of the airplane always to fly head-on into the relative wind; to yaw around as necessary to *point* in the direction in which it is actually *going*. Other ways of expressing the same thing would be: the airplane wants to "track" nicely; or: the airplane wants to line itself up with its own motion to the air. In pilot's language: it does not want to skid or slip, that is, to slither sideways through the air; but if such a slip or skid occurs it stops it by yawing into the slip. The best comparison here is the ordinary skid of an automobile, when the rear end sometimes comes around and gets beside or even in front of the front end; the directional stability of the airplane prevents it from ever doing *that* particular trick. It always keeps the airplane's tail behind its nose.

THE CONTRARINESS OF BOMBS

This seems perhaps no particular accomplishment on the part of the airplane. Yet it is a fact that a streamlined body (such as an airplane's fuselage) must be deliberately stabilized in this respect. This stabilizing is done largely by the vertical tail fin—and in general by the vertical side area of the fuselage rearward of the airplane's center of gravity. If the airplane did not have this vertical tail area, it would have a positive tendency to set itself crosswise to the Relative Wind, that is, crosswise to its own motion. That is why bombs have stabilizing finds on their tails; if they hadn't, they would tend to fall crosswise. It is also why airships and captive balloons have stabilizing fins. And it is why bullets are given a rapid rotation by the rifling of the gun. If it were not for this gyroscopic stabilizing, they would set themselves crosswise to their line of flight and execute the kind of motion called "tumbling" or "keyholing." Thus this weathercocking tendency is by no means an unimportant characteristic; the airplane would be unflyable without it. But all it does is to keep the ship's tail from skidding around and setting the airplane crosswise to its line of flight. It does not keep the airplane's flight from curving. On the contrary, it makes its correction by yawing the airplane around, and thus under some conditions it actually causes the airplane to turn.

SPIRAL STABILITY

This leaves us with an airplane that "wants" to fly at constant angle of attack and "refuses" to sideslip or slice sideways through the air. But that still doesn't answer the question: what does the airplane "want" to do concerning the *direction* of its flight? Does it want to fly straight?

Let's state the facts first and discuss the reasons later. There are reported to be some airplanes (most of them foreign) which positively prefer to fly straight and which, should some disturbance put them into a turn, will straighten out from the turn and resume more or less straight flight at least until a new disturbance puts them into a new turn. There certainly are model airplanes that behave that way, but the average airplane, as built to American specifications, does not have that sort of stability. It will not fly straight but—with the controls released—will positively go into a turn. It will then stay in the turn; and usually it will gradually steepen and tighten its turn and drop its nose and wind up in a spiral dive.

Most pilots know this but will not believe their own experience, because it does not agree with carelessly written how-to-fly books that keep stating that the airplane wants to fly straight—the motive being to impress the reader with the inherent safety of flight. When he releases the controls and his airplane thereupon begins to turn, the pilot therefore tends to blame "torque." But experimentation on a glider would disprove that.

A glider has, of course, no torque. Yet most gliders do not want to fly straight but want to go into a turn. Or perhaps the pilot blames the turning tendency upon "wing-heaviness." But experiment will show that even a perfectly trimmed airplane will go into a turn and that it will go off to the right as readily as to the left, depending entirely upon chance. Or again the pilot may blame stiff controls. In many airplanes there is so much friction in the control system that the controls, when released, will not neutralize or streamline themselves but get stuck slightly off center, thus giving the ship constant slight rudder or constant slight aileron on one side or the other. But the controls can be set to center by a slight tap on the stick or the pedals; the ship will even then still want to go into a turn.

Of course it is conceivable that a perfectly maintained airplane, perfectly trimmed and flying in smooth air, might continue to fly straight ahead indefinitely simply because it would never be disturbed, would never get that first slight deflection from straight flight which starts the spiraling. But it is also conceivable that a pencil, stood on its point, might be so perfectly balanced and so completely shielded from all disturbances that it would stand on its point indefinitely. Both things are philosophically possible; both are so highly improbable as to be practically impossible. And anyway, the stability of a thing, just like that of a man, does not consist in its being shielded from all disturbances and thus preserving a precarious balance; it consists in the ability to recover from disturbances that inevitably will occur, and to regain lost balance.

The pilot should keep that in mind when he is tempted to blame the turning tendency of an airplane on wing-heaviness or torque, rather than on its inherent spiral instability. If the airplane were stable in straight flight, if it really *wanted* to fly straight ahead, it would fly straight, or pretty nearly so, even despite such minor disturbing forces as torque, wing-heaviness, or off-center controls. Recall to mind once more, as an example, how the airplane acts in regard to speed and/or Angle of Attack—a sense in which it is truly stable. If a passenger leaves one of the rear seats and moves farther forward, you have in this sense a case that is exactly analogous to wing-heaviness in the straight-flight sense: the center of gravity has shifted, and the ship is out of balance. Does this disturbance put the airplane into a vertical dive? No, its speed-keeping tendency takes care of the disturbance; the only result of the passenger's moving forward will be a very slight nosing down and a very slight build-up in speed—and at the slightly higher speed the airplane will find a new balance. It does not continue to put its nose down farther and farther; it does not continue to build up additional speed. In the same way, if the airplane really wanted to fly straight it would resist such small disturbances as wing-heaviness or torque. It would perhaps turn a very little, but the small disturbance could never succeed in steepening the turn progressively and tightening it into a spiral dive. The plain fact is, then, that the average airplane does not want to fly straight but positively wants to turn.

It is true that not all airplanes will tighten and steepen their turning into a spiral dive. Some will very soon reach an angle of bank and an air speed, a rate of turn and a rate of descent at which a stable equilibrium is found. If such a ship starts from straight and level cruising flight at 100 m.p.h., equilibrium may be established in turning flight at, say, 20 degrees of bank and, say, 120 m.p.h. of air speed—the resulting rate of descent being perhaps 300 feet per minute. The flight path therefore will be what airport language calls a *spiral* and what should really be called *helix;* the ship will circle while descending steadily. By readjusting the trim tab or by using additional power, the loss of altitude can be stopped, and thus such a ship can be made to circle, with all controls released, as long as its fuel lasts.

Most airplanes will, however, keep increasing the angle of bank, the rate of turn, the speed, and the rate of descent. The result will be a true spiral, that is, an ever-tightening turn—combined with an ever-steepening dive. If left to its own devices long enough, such a ship will finally corkscrew down at terrific speed, banked so steeply that observers on the ground might call it slightly inverted. Because of the high speed and the tightness of turn, the *g* load becomes so fierce that airplanes have broken up in such dives. The motion superficially resembles a spin; but it differs from a spin in that no stall is involved and that the controls will function in a normal manner as soon as the pilot decides to use them.

WHAT CAUSES A SPIRAL DIVE?

Now for an attempt to sketch the causes of the airplane's spiral behavior. They are fully understood only by a few dozen men in the whole world, unfortunately not including the writer. Even most of the men who design airplanes deal with the problem of spiral stability rather by rule of thumb, by trial and error, by making the new ship similar to some existing ship which is known to behave well in this respect. Hence we are talking here, reluctantly, about something that we do not really understand, and the sketch must be sketchy indeed.

The key fact is this: If the airplane's flight is disturbed in the right-and-left sense—and it makes no difference what the disturbance is, whether a gust drops one wing or whether "torque" swings the nose to one side, or whether wing-heaviness forces one wing down, or whether a momentary sharp side gust slews the tail around—if there is any disturbance in the right-and-left sense, the airplane's stability will always respond to that disturbance in two ways, *both ways at once.* Its weathercocking stability, due to its vertical tail area, will cause it to yaw around so as to hold it pointing head on into the direction in which it is actually moving. And *at the same time,* its tendency to refuse to sideslip, due to the dihedral angle of its wings, will cause it to lift one wing and drop the other so as to regain its lateral balance. Each of these two reactions has already been discussed in detail. What matters here is that any one right-and-left disturbance of the airplane's flight will always call forth *both* reactions. For any such disturbance causes the relative wind to blow slightly crosswise at the airplane—that's why it is a disturbance; and this crossflow of the air acts on both the vertical tail and the wings' dihedral. For example, if a gust has put down the right wing and has thus started a slight sideslip to the right, the ship responds by weathercocking to the right. But at the same time it responds by picking up its right wing again, that is, by rolling to the left.

Whether the disturbance results in a change of flight direction depends, then, on whether the weathercocking response or the wing-righting response is quicker and more forceful. For example, imagine an airplane with very pronounced dihedral and an exceptionally small vertical tail fin. Many model airplanes are designed that way. If a gust throws such a ship into a slight sideslip to the right, the right wing will pick itself up before the vertical fin has had time to yaw the ship around to the right. After the disturbance has passed, the ship will thus still fly substantially in the old direction. By contrast, imagine a ship that has no dihedral at all and an exceptionally big vertical tail fin. Such an airplane will respond to the same disturbance almost purely by yawing, and if it recovers at all (which it does not, for reasons to be shown later), it will recover on an entirely new heading. It is interesting to note the trouble maker; the vertical fin, whose purpose many student pilots think is to keep the airplane flying straight, is actually what makes it turn!

TIMING

Which of the two reactions predominates in any given airplane depends not only simply on the corrective forces set up, but also on the quickness with which they act; it depends, therefore, not merely on the degree of dihedral and the square foot area of the vertical fin, but also on such factors as tail length and wing span; generally speaking, a close-coupled tail fin, carried on a short tail, is quicker acting, despite its shorter leverage, than a tail fin that is carried on a long tail. The concept of "damping" becomes important here—crudely speaking, the fact that an airplane's wings will not roll, and its tail will not slew around in yaw, with more than a certain quickness, because the density of the air won't let it, much as the density of the water won't allow you to move your hand under water as quickly as you can move it through air. But at this point, the whole discussion becomes too abstruse to continue. The trouble is, the airplane won't hold still while we think about it; the forces acting on it are hard enough to understand; the time sequences in which they act, the rhythms, as it were, of the airplane's behavior, are even harder to grasp, and they seem to be the essence of the thing. The whole process becomes impossible to describe without lapsing into mathematics.

Perhaps these sour grapes are the designers' business anyway rather than the pilots'. For our purposes here, it may be more useful to go back to things that concern the pilot more directly.

Why is it, a pilot might ask, that all ships are not designed to be spirally stable? If model airplanes can be made to prefer straight flight, why not real airplanes? Designers give a double answer to that one. In the first place, they say spiral stability is not particularly worth having. We don't propose to go to sleep in our airplanes. The idea of stability is not that the airplane should actually control itself, but that its various willingnesses and unwillingnesses should help the pilot in his work of controlling the airplane. And as long as an airplane has those three basic kinds of stability—speed-keeping tendency, weathercocking tendency, and tendency to right itself from a sideslip— it gives the pilot all the help he needs. Moreover, designers say, a spirally stable airplane is unpleasant to fly in rough air. It is wallowy and unsteady and wears you out.

Be that as it may. A question that concerns the pilot more directly is this: How does a ship, once started on a turn, steepen and tighten that turn and work itself into a spiral dive? The reason why this concerns the pilot is that any tendencies that exist in a turn with controls released will of course be noticeable also when the pilot "flies" the turn.

THE BACKGROUND OF A TURN

Boiled down to the simplest terms, here is what happens: In straight and level flight, the right wing drops slightly. The airplane thereupon begins to slip slightly to the right. To the resulting crossflow of air, the dihedral responds by trying to pick the right wing up again. But the tail fin responds by yawing the ship slightly to the right; and if the ship is spirally unstable, the tail fin forces the ship around before the dihedral has had time to pick up the wing.

Now this yawing ("turning"), however slight, introduces an *overbanking tendency;* while the ship yaws around, its left wing is temporarily being thrown forward, going through the air a little faster, and thus getting more lift; while the right wing at the same time is being temporarily retarded, thus getting less lift. The result is that the left wing tends to go up slightly while the right wing drops; this overbanking tendency nullifies the attempt of the dihedral to right the ship. Thus the right wing stays down, or goes down still more, and the slight slip to the right, which triggered off the whole process in the first place, is continued. To this continued rightward slipping, the dihedral responds *again* by attempting to raise the ship's right wing and thus to stop the slipping. But *again* the vertical fin responds to this same slipping by trying to yaw the ship around to the right, and again the yawing response is quicker and predominates over the wing-righting response. This new yaw-around then results in a new overbanking tendency, and thus in new sideslip to the right; and thus the whole process continues itself.

Let it continue long enough, and you get a steeper and steeper bank, a tighter and tighter turn. As the bank increases and the turn tightens, the "*g* load" due to centrifugal force builds up; and, as the ship feels this additional load, it responds by dropping the nose and picking up speed. For—as has been shown in our discussion of longitudinal stability—an airplane has a built-in tendency to keep itself at constant Angle of Attack, and when it is loaded up with additional "weight" (even if that weight is only the apparent weight due to centrifugal force), it can maintain its Angle of Attack only by picking up additional speed. Thus a dive combines with the spiral to make a spiral dive.

As soon as the airplane's bank has reached an appreciable degree, there is also another effect that forces its nose earthward; with the ship thus banked, the earth is on its right side, the sky on its left side. And thus, when its vertical fin yaws it "around" to "the right," it now really turns it earthward; the effect is similar to the effect of holding constant "bottom" rudder during a steep turn.

That, then, is what the airplane "wants" to do once it is in a turn; it "wants" to overbank, it "wants" to sideslip inward, toward the low wing, and it "wants" to put its nose down. It "wants" to do these things, of course, also when the pilot is on the controls. And the disposition of the controls during a turn and the pressures the pilot exerts on stick and rudder during a turn are nothing but an effort to block the unstable, ill-behaved intentions of the airplane—aileron to the high side to block the overbanking tendency and keep the bank from steepening and sideslip from developing; back stick, to force the airplane to fly at higher Angle of Attack, and thus keep it from speeding up as it feels the additional "weight" of the *g* load; and, at least in simple principle, no foot pressure on the rudder because the tail fin is active enough anyway or really overactive in turning the airplane.

But the flying of a turn will be discussed in a later chapter. Right here, something else should be made clear.

A WARNING

That the average airplane really *wants* to spiral dive is the cause of an effect that has cost many a life and should be more clearly appreciated by many people, especially by people on the "outskirts" of aviation—student pilots, teachers of aeronautics in high schools, aviation "officials" of various sorts. The exact causal connection is too involved to trace here, but the effect is this: *If an untutored person tries to fly an airplane* and uses the controls in the manner that seems most "natural" to him, responding most energetically and most quickly to those disturbances that "naturally" impinge most sharply on his consciousness, *the flight will almost certainly end in a spiral dive and a crash.*

This goes in the first place for any attempt on the part of pilots to fly "on instruments" without proper instruction; the spiral dive comes usually after a very few minutes. It goes, next, for any attempt to fly a conventional airplane without previous instruction—and it makes no difference how cool-headed or intelligent the would-be flier is. It goes, therefore, especially for boys who want to "steal" an airplane and fly it without previous instruction. The probabilities are slightly less unfavorable in such a case than they are for a contact pilot's instrument flying, but they are still disastrously unfavorable. Therefore it ought to be made common knowledge that the untutored attempt at flying will end in a spiral dive as surely as a first attempt at skiing will end in a heavy sitdown.

And this warning goes very much for those enthusiasts who still advocate large-scale flight training, possibly of high-school kids, in primary gliders. In such ships, the flier is on his own from the very first hop on. What's worse, he begins to learn his "flying" while skimming on the ground, when the airplane's responses to the controls, and its inherent tendencies are all falsified by this fact of ground contact; when the pupil finally gets into the air he knows really nothing about the real tendencies of his ship and the real effects of his controls. Experience shows that in fact the typical glider accident is not the stall or the spin but a spiral dive. Disregarding the glider pilots properly trained during the war, and disregarding a few very able pilots, one might almost say that few gliders are flown more than a couple of hours total flying time before ship's and pilot's career is interrupted, to say the least, by a spiral dive crash. This is not because gliders are necessarily more unstable than power planes; but because in the past, many glider pilots were self-trained by the hop-skip-and-jump method in very short jumps and were thus bound to spiral-dive helplessly the first time that the combination of flight circumstances invited a spiral dive.

Chapter 8

THAT THING CALLED TORQUE

OF THE several stupid things about the present airplane, the most stupid is that thing called *torque*, that insistent tendency of the airplane to turn to the left which forces the pilot to hold some right rudder pressure throughout most maneuvers merely to keep the airplane straight! The effect is clearly due to the power plant—for it shows strongest at wide-open throttle, and it is absent in a power-off glide. It has something to do not only with the speed at which the engine is turning, but also with how hard it is pulling. At any given throttle setting it is the stronger the more slowly the airplane is flying—for example, take-off runs, climbs, and steep turns require right rudder pressure; dives (the other extreme) sometimes require left rudder pressure merely to keep straight.

Now it is bad enough that an airplane has a rudder at all, seeing that "the only purpose of the rudder is to cover up the mistakes of the designers," but it is even worse that this annoying effect, torque, makes it impossible to use the rudder in a consistent, simple, logical fashion; because of torque, we fly straight holding right rudder, we dive straight holding left rudder, and we sometimes go around left turns holding right rudder! And although we know it is one of the worst sins of the pilot to hold rudder during a turn, we do hold right rudder during a right turn—because of torque. Torque messes up our whole handling of the controls and is especially annoying to beginners; since correct footwork on the rudder is the most difficult part of elementary flying anyway, torque makes learning to fly doubly difficult.

To add to the confusion, this left-turning tendency is misnamed. It is not even due to the effect that most pilots blame for it; "torque" is not really torque at all!

But let's see.

Three theories have currency around the airports concerning the cause of this left-turning tendency. All three have their points and are worth understanding; but all three are not equally important. Only one really hits the spot. It may be best to clear away the less important ones first.

THE SPINNING TOP

Some old-timer might tell you that what makes the airplane turn to the left is a "gyroscopic" effect. This is true—sometimes. The engine and propeller of an airplane are definitely a "gyroscope," that is, they are a rapidly spinning body that has considerable mass. If such a spinning mass is disturbed so as to change its axis of rotation (in this case so as to swing the ship's nose up or down or to one side) the spinning body does not respond to the disturbing force as a non-spinning body would respond, that is, by simply obeying that force. Instead, it "pre-cesses" sideways, that is, evades the force at right angles. In the case of an airplane with the usual clockwise-turning propeller, pulling the nose up produces a gyroscopic effect that swerves the nose to the right. Kicking the nose to the right by rudder produces a gyroscopic effect that makes the ship nose down. Pushing the stick forward makes the nose swerve to the left. Kicking the nose to the left by rudder makes it come up!

Old-timers knew that! Many pursuit ships during the First World War were powered by rotary engines—the crankshaft was fixed, and cylinders and crankcase rotated around it, along with the propeller. This made the spinning mass comparatively large—and at the same time the ships themselves were light, quick, and jumpy. Hence the gyroscopic effects were marked, and overshadowed all the more normal responses of the ship; the pilot's handling of stick and rudder in such a ship would have been completely puzzling to an observer unaware of those "precession" effects. So marked were these effects that the enemy counted on them in combat; a rotary-engined pursuit, attacked from the left, and whipping around to the left to meet the attack was likely to climb or even stall. If attacked from the right and thus forced to make a quick turn to the right, it was likely involuntarily to have its nose forced down, thus losing altitude.

On modern airplanes with normal engines and better stability characteristics, the precession effect is but faintly noticeable in normal maneuvering. But it does explain some peculiarities of behavior that would otherwise be quite puzzling. Why, for example, do many airplanes snap-roll more easily to the right than to the left? When at the start of the maneuver the pilot pulls the stick back and the nose swings up suddenly, gyroscopic force swerves the nose to the right, thus adding its effect to that of the pilot's right rudder and thus helping the snap roll along. In a snap roll to the left, the gyroscopic force is, of course, the same, because it is due to the upswing of the nose; but now, in a snap roll to the left, its force counteracts the pilot's left rudder, and this results in a more sluggish roll.

Or, a pilot may notice that, in practicing steep figure eights "around" two pylons—a maneuver that requires a rather whipping entry into a steep turn—he loses altitude on the right turns although he is not conscious of any control errors. The trouble is probably due to gyroscopic precession, which forces his nose down and to the right and must be counteracted by both stick and rudder.

TROUBLES IN THE TAKE-OFF RUN

The take-off is another occasion when gyroscopic precession can become quite noticeable—or even dangerous. There is a left-turning tendency during the take-off run anyway—for other reasons. But, as the pilot raises the tail, the nose is bound to swerve to the left with some extra impulse! If the pilot pops the tail up suddenly, this sudden swerve to the left can be quite pronounced. It may help a student to understand this effect and how it is actually a result of raising the tail. It will help him to anticipate the need for foot action to counter the swerve by some extra pressure on the right rudder, timed and proportioned just right. It will also help him to refrain from raising the tail too roughly.

This effect can become dangerous on twin-engined ships. If such a ship has a single rudder, located so that the slipstreams of both propellers pass it by, rudder control during the early stages of the take-off is weak; if the tail is raised early in the run and suddenly, the combined precessions of the two power plants, combined with the weakness of the rudder and with the directional instability of the conventional landing gear, may produce a vicious ground loop right then and there. Among pilots, this effect is sometimes described as "stalling of the tail." The tail surfaces are thought to be stalling because of lack of speed and the pilot's attempt to lift the tail too hard; and in stalling, they are thought to swing sideways. The precession effect seems a more likely explanation.

There definitely are occasions, then, when the gyroscopic effect is important even in present-day ships. It does explain all sorts of things. But you will have noticed that it does not explain that thing usually called torque, the ordinary tendency of an airplane to turn left under the influence of its power plant.

That tendency is present even in perfectly straight flight during a straight climb, for example, when the nose is not swerving in any sense, and gyroscopic precession can't account for it, because precession occurs only when there is a swerve.

Some other explanation is clearly needed.

ACTION AND REACTION

The explanation now most popular among flight instructors is the one that has given the effect its name, torque. The engine, in turning the propeller one way 'round, turns itself (and thus the whole airplane) the other way 'round, since "action" is accompanied by "equal and opposite reaction." The propeller rolls clockwise, and the ship tends to roll counterclockwise, that is, to the left.

This effect—so the torque theory goes on—is continually forcing the left wing down. The designer therefore puts the left wing at a slightly higher angle of incidence than the right wing, thus giving it some extra lift. But this gives it also some extra drag, and that extra drag would cause the airplane to yaw to the left. To keep the airplane from yawing the designer then sets the vertical tail fin at a slight angle, so that it has the same effect as if the pilot were continually holding right rudder. Thus the designer achieves an airplane that wants neither to roll nor to yaw. But he gets this perfect balance of forces only for one combination of flying speed and throttle setting. And therefore—the theory allows you to infer—the ship will turn to the left when flown at a slower speed and wider open throttle—as for example in a climb. It will turn to the right when flown at faster speed with the throttle less wide open.

This theory has flaws—which would not matter if it succeeded in explaining how the airplane behaves. But it does not; sooner or later it will leave the pilot puzzled.

It is true that the engine, in turning the propeller one way 'round turns itself the other way 'round. This effect is important, too—in some types of ship and in certain flight situations. But it is not important in ordinary trainers or small private and commercial airplanes or in ordinary flight. For this propeller torque is too feeble compared with the airplane's own powerful inherent stability, its tendency to keep itself from rolling. Torque acts on a short lever arm—the propeller. Stability acts on a long lever arm—the wings. The propeller is so short and the wings are so long that any twisting force on the propeller cannot overpower the wings' tendency to hold themselves level when in straight flight.

Next, it is probably true that the left wing is in most airplanes set at a higher angle of incidence than the right wing; but for most trainers and other low-powered airplanes it is not true that the designer actually sets it thus. The test pilot simply takes the ship up, checks it for wing-heaviness (for which there are many possible causes in addition to torque); the rigging crew then adjusts it. And it may come out of this adjustment process with this slight difference in wing incidence rigged into its wings.

At any rate, all this doesn't add up right. For this propeller torque will tend to *roll* the ship. Hence in flight situations where this torque is strong, we should notice a tendency to *roll*, not a tendency to yaw. We should have to hold right aileron to counteract it, not right rudder! In short this torque effect may exist, but it is not what bothers us in the ordinary airplane. The "torque" theory does not explain the thing we seek to explain, what makes our airplanes turn to the left.

TORQUE CAN BE DANGEROUS

This does not mean that torque is always altogether negligible. And since you have plowed through the subject so far, you might stop here for a moment and consider certain situations in which this true torque does become important and can even become dangerous.

Obviously, it would be most important in airplanes of short wing span and big powerful propellers. Pilots report how on his first hop in a Spitfire an inexperienced man is likely to stagger off left wing low because of the tremendous torque of a suddenly accelerated engine. *Accelerated*, for this propeller torque is most forceful not (as many pilots think) when the engine is running at wide-open throttle, but at the moment when you accelerate your engine, that is, when you gun it! On the other hand the force that resists it—an airplane's lateral stability—fades out as the airplane slows up; finally, when the airplane stalls, it turns into instability. Thus there is one situation when torque can become really dangerous: when you are stalled or nearly stalled, and then gun your engine suddenly. The torque can then roll you over. This is not a serious danger, however, in trainers and other low-powered ships; it should not keep the student from using a blast of the throttle freely on such a ship to get out of trouble. But in hot, short-span, powerful ships, the throttle must be handled with caution.

If an airplane is very powerful and has perhaps a narrow landing gear, torque can also upset it while it stands on the ground—if the pilot or mechanic suddenly guns the engine.

THE SPIRALING SLIPSTREAM

So much for torque.

The true reason for the airplane's left-turning tendency is not "torque" at all and is little understood among pilots. It is very simple; the propeller washes back a stream of air which is not in straight

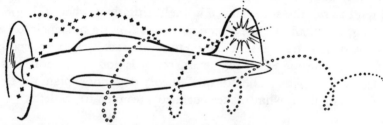

What pilots call "torque" is caused mostly by the position of the tail fin with respect to the spiraling propeller slipstream. Note how air particle + hits the tail and shoves it to the right, while air particle O passes unhindered underneath the tail, failing to produce a balancing counterforce to the left (spiraling greatly exaggerated).

backward flow but spirals as it streams to the rear. Because of its spiraling motion, this air stream hits the vertical fin at a slight angle, thus pushing the tail to the right and yawing the nose to the left. And that's all!

To get the idea of this spiraling, suppose you follow the path of one certain particle of this air. It is caught by the propeller when it is at the right-hand side of the ship's nose. The impact of the propeller sends it backward and downward along the right side of the fuselage; it crosses underneath the pilot seat to the left side of the fuselage; it flows upward and backward along the left side of the tail; and it is then ready to cross over the top of the tail back to the right side; but here it finds its path blocked by the tail fin; and its impact on the tail fin shoves the tail fin to the right.

At the same time another air particle is caught by the propeller somewhere on the left-hand side of the nose. The impact sends it speeding backward along the left side of the fuselage and slightly upward. It crosses over the top of the cockpit and flows downward and backward along the right-hand side of the tail. It then is ready to cross underneath the tail back to the left side. And here is the trouble; there is no tail fin down there to block its path. Thus it exerts no force but flows away unhindered.

The net result of these two particles' passage is that the tail has received one shove to the right but no shove to the left. And hence it moves to the right, yawing the airplane to the left.

This description exaggerates the tightness of the spiral. In cruising flight the slipstream actually spirals so little that one might almost call it straight. Designers usually set the tail fin at a slight angle so as to line it up with the actual direction of the air flow at the tail in cruising flight; and that is the true reason for offsetting the fin—not to make a force, but to keep a force from being made!

In slower flight at wider open throttle, when the propeller blades meet the air at higher Angle of Attack, the spiraling is much more pronounced—though still not nearly so tight as our description makes it out. As the tightness of the spiraling increases, the sideways shove on the tail fin increases. If the tail fin is not sufficiently offset, the slipstream hits its side and a left-turning tendency appears. The pilot must hold right rudder.

In gliding flight, when there is no slipstream, there is of course no spiraling of air; the flow over the tail is then straight. The offset of the vertical fin is then unnecessary and in fact yaws the ship to the right: the pilot must hold slight left rudder. This becomes apparent to him, however, only if the glide is fast enough to be called a dive.

You would expect the airplane to show a very pronounced left-turning tendency while approaching a power stall. It does. But when the "break" of the stall has occurred, the left-turning tendency temporarily disappears in some airplanes even though the engine is still wide open, and the ship is still slowed up, and the pilot expects to have to hold much right rudder. This puzzles many pilots. What happens is that, after the stall has occurred, the airplane sinks through so fast that the slipstream fails to hit the tail. It takes an appreciable time for the air particle to complete the journey from the propeller to the tail; and by the time it gets there, the whole airplane, including the tail, has fallen out of its way! Hence the air flow over the tail is straight, and for the moment there is no left-turning tendency.

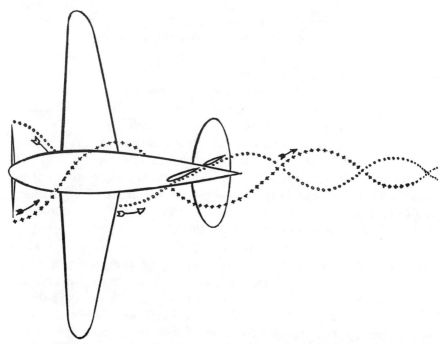

To keep the spiraling slipstream from yawing the airplane, the designer sets the tail fin at an angle. Thus the airplane does not yaw to the left so badly with power on; on the other hand, it wants to yaw to the right with power off (spiraling and offset greatly exaggerated).

Could this left-turning tendency be abolished altogether? It could be done by putting as much vertical tail fin underneath the tail as there is atop the tail; this would block the path of our second imaginary air particle and would equalize the shoving effects of the slipstream. The reason it isn't done is that it would interfere with the conventional landing gear which calls for a tail wheel in just the location where the lower tail fin would have to be.

Now modern safety airplanes have no rudder and must therefore somehow get rid of the left-turning tendency. In such airplanes the problem is solved by providing a double tail, that is, one small vertical fin at each tip of the horizontal stabilizer. This takes the vertical tail area almost entirely out of the slipstream, out into undisturbed straight-flowing air; the slipstream passes between the two fins. What

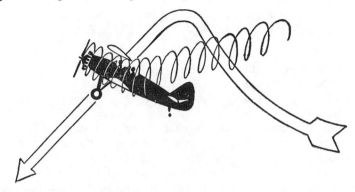

Why "torque" may temporarily fade out in a power stall: the airplane temporarily sinks so fast—just before dropping its nose—that the slipstream does not reach the tail.

little then remains of left-turning tendency is compensated for by mounting the engine at a slight angle, so that it pulls just a shade to the right. There is no reason why those two tricks—the H tail and the slanted engine—could not be used in conventional airplanes as well, except that too many pilots feel that an airplane isn't a "real airplane" unless it faithfully reproduces all the traditional vices all over again.

PART III

THE CONTROLS

The controls of the airplane become the harder to understand the more you fly. Those first beautifully simple ideas—push forward on the stick if you want to go down, pull back if you want to go up, and so forth, dissolve during the very first hours of actual flight experience—in the glide, for instance, you go down although the stick is held well back. In the spin, you go down quite clearly *because* the stick is all the way back! The more sophisticated a pilot becomes, the more he discovers that his controls have a wide variety of effects, and all sorts of by-effects, and sometimes the by-effects seem more important than the main effects. And he finds out that, in some situations, the controls actually reverse their usual effects and deliver results exactly contrary to those expected!

Hence, experienced pilots usually don't care to formulate a theory of the controls; they tend to say that it all depends—you've got to know how to fly, that's all. And of course, the experienced pilot needs no such theory. By the time he has flown enough to formulate one, he is no longer puzzled about the controls. He has the feel of them.

It is the student who needs a theory (that is, a clear mental image) of his controls. Much of the difficulty of flying, much of the wasteful inefficiency of primary flight instruction is due simply to the fact that the student does not know really *what* he is doing when he moves his stick, his throttle, or his pedals. Think how absurd it is: piloting is done entirely by moving certain levers in the cockpit, but the student pilot does not know what each lever *is* for!

He does not know because he is not told. He is not told because the instructor fears that a thorough discussion of control effects would only confuse him. "Control analysis," as the experienced pilot understands it, is an enormously complicated lore, an almost endless list of small tricks and odd complications, unexpected effects and disturbing by-effects. It is the accumulated experience of hundreds of hours of flying, of thousands of observations—most of them made unconsciously. The student is indeed better off if he is simply allowed to gather this knowledge for himself—even though it will take him, too, hundreds of hours of flying to do so.

But still, the student ought to have a sound explanation of the controls. He is usually too confused and overawed to ask for one, outright; but he feels the need. Probably every student pilot has felt, with an acute sense of frustration, that important things, helpful things, key things were being left unsaid;

and he has wished that he had sense enough to ask his instructor some key question. Almost certainly this is one of the questions that he would ask, after a couple of hours of dual instruction after his first oversimplified ideas have evaporated: "Well then, what *is* this lever, this pedal, this wheel really for?"

Fortunately, he *can* be answered. The real trouble with "control analysis," as current at our airports and flying schools, is not that it goes too far. It does not go far enough. It is complicated not because it is analytical, but because it is not analytical enough. It remains a lore, a body of assorted experiences; the principles are not distilled out. If control analysis is carried far enough, it becomes simple and most helpful to the student.

For actually the controls of the airplane are simple and straightforward. Actually there are no reversals of control effect, no exceptions, no tricks. Each control does a certain thing, does that thing only, does it always and faithfully; but the thing it does is not what you first thought it was. To understand your controls, then, all you have to do is to find, understand, and clearly envisage the one thing that each control does.

This will be attempted in the following three chapters.

Chapter 9

THE FLIPPERS AND THE THROTTLE

YOU know how the controls are labeled on an alarm clock or on a kitchen stove. There are arrows marked "on" and "off," "slow" or "fast," "hot" or "cold," telling the customer exactly which way to move what control in order to get what result. Well—here is an idea for one of those unflyable days at the airport; how would you label an airplane's elevator and throttle?

Your kid brother, knowing what every boy knows about flying, will call this one easy. He will label the throttle "fast" and "slow." Does not the throttle do to the airplane's motor exactly what the gas pedal does to an automobile motor? And he will label the stick "up" and "down." Does not pulling back on the stick make the airplane go up? It is only common sense.

THE ELEVATOR DOESN'T ELEVATE

Unfortunately, though, the present conventional airplane is not a common-sense contraption; and this labeling of its controls is wrong. It is wrong not only "in theory"; it is wrong also in practice. It is dead wrong; if you really did try to use the controls that way, you would kill yourself. Most fatal airplane crashes happen precisely because the pilot has the controls so labeled in his mind and tries to "elevate" himself, or at least to hold himself up, by pulling back desperately on the so-called "elevator." An airplane will not go up, nor will it stay up, simply because the pilot pulls the stick back. In fact, in all the more critical situations of flight, that is, when climbing steeply, or gliding slowly, or in turning flight, it is all too likely to do exactly the contrary! In the glide, for instance, the farther the stick is held back, the more steeply downward will the flight path be—even though the nose may not point down steeply. In a stall or a spin, the ship is dropping precisely just because the stick is held too far back! As for the throttle's being a speed control, the fact is that you can stall with the throttle wide open! Most fatal stall or spin accidents do occur with the engine running nicely at cruising throttle! In fact, when the throttle is open and the propeller blast is hitting the tail, the average airplane actually wants to fly more slowly, and will stall more readily, than with power off! In short, the elevator does not make the airplane go up or down, and the throttle does not make it go slow or fast, and your kid brother is wrong.

A DISTILLATE OF WISDOM

Some old-time pilot, if you asked him, would say that of course that's so. Of course the controls cannot be labeled in terms of the flight path at all. Of course pointing an airplane's nose up will not necessarily make it go up. Why—all that is perhaps the most fundamental bit of wisdom an old pilot distills out of years of flying experience—this, that pulling the stick back won't necessarily make the airplane go up, and then the other, that kicking the rudder is not the way to make the airplane turn. If you can keep those two firmly in mind, you'll be all right.

But if you should press the old-time pilot for a positive answer—what does the stick do, and what results does the throttle get—he would probably say that the stick should be labeled "nose-up" and "nose-down"; with the terms "up" and "down" referring not to the sky and the ground, but to the pilot himself; pulling the stick back will always pull the nose "toward" the pilot, pushing it forward will always push the nose "down," that is, "away" from the pilot. In fact this is now the standard explanation of control effects that instructors usually give to a student before his first flight.

It is not quite good enough. It works fairly well in normal flight when it doesn't matter so much whether you exactly understand your controls; but it lets you down in the emergencies, at the exact time when a clear understanding is the most important. It fails, for one thing, to cover the case of the stall; by holding the stick back too far the pilot actually makes the nose fall down, "away" from him! And it fails to cover the case of the spin; the harder the pilot pulls back the stick trying to get that nose "up," "toward" him, the more obstinately does the nose stay down, "away" from him.

THE ANGLE OF ATTACK CONTROL

Thus cornered, your old-time pilot probably will say that the controls simply cannot be labeled in any straightforward fashion; you've got to know how to fly—that's all.

Now if you asked an aeronautical engineer, he could tell you. He would mumble something about Angle of Attack; engineers always do! What he would say would be this: the elevator determines the Angle of Attack at which the airplane will fly, and this is really what the elevator is for: it is the airplane's Angle of Attack control. At any moment, the airplane will fly at the Angle of Attack which the pilot prescribes for it by holding the stick wherever he is holding it.

But because "Angle of Attack" sounds so theoretical to you, you would probably stop listening to the engineer; pilots always do! This is a pity. The airplane is a machine, and whatever its designers have to say about it should be of interest to the pilot. If the engineer comes back every time to the concept of Angle of Attack, he does so for good reasons. Angle of Attack is the central fact of all flying; and a pilot who refuses to think in terms of Angle of Attack simply does not understand the airplane—even if he knows the whole rigmarole that commonly passes as Theory of Flight.

Let's make a quick attempt to show what the engineer means. To him the elevator is nothing but a quickly adjustable stabilizer; or rather, a quickly adjustable extension of the horizontal stabilizer. When the pilot pulls the stick back and thus deflects the elevator upward, the net result is the same as if he had simply changed the angle at which the horizontal stabilizer is attached to the airplane's tail: the airplane then assumes a different Angle of Attack. And as long as the airplane is held in that position, the airplane will continue to fly at that new Angle of Attack.

Thus the elevator does nothing that could not be done equally well by the stabilizer. It is not strictly a steering control at all but is simply a control which fixes flight condition. This may sound somewhat abstruse. You will see the difference if you compare the airplane's elevator to an automobile's steering wheel; not as to *what* they do, but as to the *manner* of their functioning. As long as the steering wheel is turned to the right, the car will keep turning to the right. The car will not go straight again when the steering wheel is turned back to neutral. The elevator is quite essentially different. If the stick is pulled back, say 2 inches, the nose will swing upward as long as the stick is held back. It will presently stop this upward swing and then ride steadily in the new position: the airplane is stabilized at a new Angle of Attack.

One can therefore say with only a little exaggeration that an airplane essentially needs no movable flippers, needs no fore-and-aft action of the stick. All it needs is some device by which the angle of the stabilizer can be changed. Our conventional elevator arrangement is simply a convenient and quick way of accomplishing this change of tail angle. Some other way might work just as well. For instance, in the old airplanes, the stabilizer itself was usually adjustable, by means of a hand crank in the cockpit (like the hand crank that works the trim tab in modern ships). Those old ships could be flown quite nicely simply by working that hand crank and leaving the stick alone. In fact, it may be that our present elevator is much too quick-acting a method of changing the tail angle and thus changing the ship's Angle of Attack—that the pilot could do less harm to himself with some slower acting, more restricted device.

THE ELEVATOR IS THE MUSH CONTROL!

If you want to understand your elevator, then, keep clearly in mind that moving the "flippers" is virtually the same thing as setting the horizontal tail fin at a different angle. With this in mind, read once more what has been explained in Chapter 7 concerning the horizontal tail and just how it determines the airplane's Angle of Attack (and thus its speed). Look once more through the illustrations of the various "gaits" at which an airplane can fly: note that in each case it is the angle at which the flippers are deflected that determines the Angle of Attack at which the airplane flies.

It is not true, then, that pulling the stick back makes the airplane go up. In most forms of descent, as a matter of fact, the stick is well back and the flippers are deflected upward. The airplane goes up or down because of the way the throttle is handled. It must be admitted that the elevator has an *indirect* up-and-down effect. When you pull the stick back and thereby force the airplane into a different flight condition, some throttle setting which previously had been barely sufficient to produce level flight may then suddenly become sufficient to produce climb: because, generally speaking, the airplane flies on less power when flying at the higher Angle of Attack. But this connection between the elevator and climb and descent is too indirect and unreliable. What a pilot should know first of all about his controls is: which effect will this lever give me—*always* and *invariably* give me?

In answer to that question, then, one may say that the elevator is the mush control! The farther you hold it back the larger will be the Angle of Attack at which your airplane will fly. And as we have seen, "flying at an appreciable Angle of Attack" is called, in pilots' language, simply "mushing." We can climb, descend, or fly level at small Angle of Attack or at large angle—that is, "mushing hardly at all" or "mushing heavily." The elevator simply determines just how heavily we mush.

All this, it must be admitted, will remain meaningless to the reader who has not yet grasped the general idea of Angle of Attack. Since this idea is a rather strange one, a first reading of this book may not have made it entirely clear. Fortunately, however, this need not keep the reader from understanding what he is really doing when he moves his stick backward or forward.

Fortunately it is possible to translate the engineer's terms into the more familiar terms of "speed" and "lift." The result is then applicable only to steady flight, including dives, glides, and climbs, and practically also to the shallower turns; it is not strictly applicable to steep turns or to pull-outs from dives. But even so, it is useful understanding.

For it turns out that the engineer would label the throttle "up" and "down" and that he would label the stick "fast" and "slow."

THE STICK IS THE SPEED CONTROL

And he isn't just trying to be difficult. It is simply the truth. Those simply are the results your stick and your throttle will get for you— the only results. If you hold the stick in a certain position, forward or backward, you thereby force the airplane to fly at a certain speed; a correctly behaved airplane will then fly at that certain speed regardless of the amount of power used. If you feed it a lot of power, it will climb at that speed; if you feed it less power, it will fly level at that speed. If you cut the power entirely, it will glide at that speed.

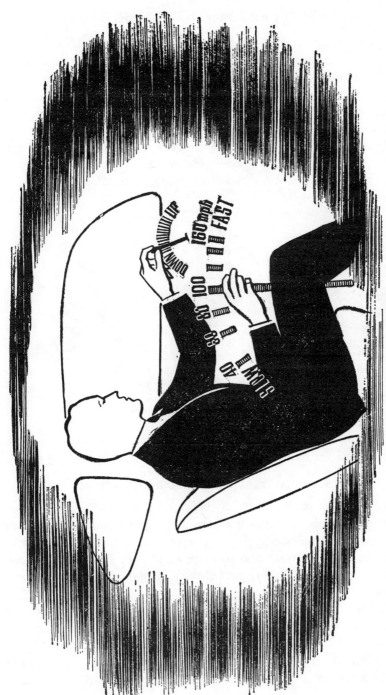

Stick and Throttle. The so-called "elevator" is really the airplane's speed control, the throttle is really its up-and-down control. This is hard to believe but is one of the keys to the art of piloting

If you put the stick back far enough the airplane will be forced to fly at stalling speed, and its flight will then take the form of a stall or a spin—regardless of how much or how little power you use. If you put the stick well forward, you force the airplane to fly very fast; and, in order to get and maintain the speed you prescribe for it, the airplane will dive.

If in fast flight you suddenly pull the stick back, the airplane, in an attempt to reduce its speed quickly to the desired figure, will balloon upward; but this upward response of the airplane is merely

The airplane's real *up-and-down control* is its throttle. The pilot here gets climb, level flight, and descent simply by changing the throttle setting. The stick is in the same position throughout, and if the airplane's stability is flawless, the air speed will be the same in all three flight conditions!

coincidental and can't be relied upon entirely; the ship's main response, and the one that can be relied upon, is the slowing up.

YOUR REAL ELEVATOR IS YOUR THROTTLE!

As for going up, there is really only one thing that will ever make an airplane go up, and that is engine power; you can make it go up temporarily by ballooning it with your elevator, but only engine power can make it go up and stay up; and hence your up-and-down control is the throttle. If any part of the airplane deserves the name of "elevator" it is the throttle.

All this is not nearly so "theoretical" as it sounds. You can demonstrate to yourself that this is really the way an airplane's controls fly. In some airplanes, certain effects (to be discussed presently) will make the demonstration a bit unconvincing. A particularly suitable airplane is the side-by-side Taylorcraft—partly because it behaves especially well in this respect, and partly because the control arrangement makes it easy to arrest the elevator in any desired position; simply pinch the control shaft with your fingers where it comes out of the instrument board. Your other hand can then still work the

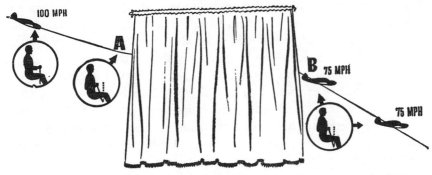

The "elevator" is really the speed control. At *A* the pilot pulls the stick back. At *B* the airplane has settled down to slower flight. Behind the curtain something happened which keeps many students from understanding the true role of the elevator.

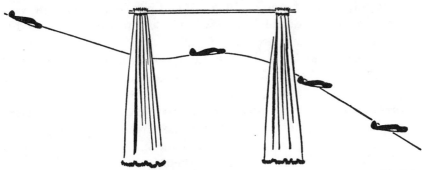

When the pilot reduces the speed there is a *temporary* upward deflection of the flight path. But it is only temporary and is often followed (as in the case shown here) by a downward deflection, so that the net result of pulling the stick back may be to make the airplane go down. The only reliable, lasting effect of the elevator is its effect on speed.

wheel for aileron control, but, with the elevator thus arrested, you can vary your throttle at will, and you will see that *the throttle is what gets you up or down*, and that *the so-called "elevator" controls the speed.*

IT'S TIME-TESTED OLD STUFF

This is also accepted and conscious practice in instrument flying. In instrument flying, the air speed is not regulated, as your kid brother would assume, by using the throttle. It is regulated by using the elevator; if the air-speed indicator reads too low, for example, you work the stick and coax it to the desired reading. You are thereby bound to lose some altitude, but that is not your immediate concern; it is taken care of later as a separate operation in the instrument pilot's 1-2-3 procedure of flying. Climb and descent is regulated, not as your kid brother would think, by the elevator, but by using the throttle. If the rate-of-climb indicator shows that you are going down, you make it read zero again by advancing the throttle.

When this is explained to a pilot at the beginning of an instrument flight course, it usually strikes him as an odd and hind-side-to procedure, peculiar to instrument flying. Actually it is neither hind-side-to, nor is it peculiar to instrument flying; it is the way he has really been using the controls all along!

The landing approach with power on—the kind of approach usually made in heavy airplanes—is another perfectly "practical" maneuver in which the elevator *must* be used as the speed control and the throttle as the up-and-down control. It just simply does not work unless you use the controls in this fashion. But this will be discussed in detail in the chapter concerning the glide.

In thinking of the elevator as the speed control, we must again make allowance for the fact that a change of speed may cause the airplane to climb or to descend; that therefore the elevator has an indirect (and unreliable) effect also in the up-and-down sense. Generally speaking, the more slowly you fly an airplane the less power it requires simply to push itself through the air; hence if you set an airplane's throttle for level flight and then slow the airplane up, there will suddenly be a margin of excess power which the airplane uses up by climbing. All this will be made a little clearer in Chapter 19, The Working Speeds of the Airplane.

EXCEPTIONS AND COMPLICATIONS

Now for some important complications. What has just been stated is literally true only in an airplane of perfect control characteristics. Unfortunately, most airplanes are not perfect. In most airplanes, the tail surfaces are exposed to the propeller blast, and thus the stick is more powerful when the throttle is open. Also, the thrust of the power plant may have the effect of nosing the airplane up.

Hence, with the stick in a given back position, the average airplane *will actually slow down when the throttle is opened wider* (nosing up higher as it does so). And there are certain stick positions (pretty far back but not all the way back) which with power off will not slow the airplane up to stalling speed, while with power on the same stick position will stall the airplane. This is why so many fatal stall-and-spin accidents occur with the engine running at cruising power. This is also why it helps to use a blast of power if you want to achieve a clean sharp stall. In some airplanes, you may actually have trouble getting the airplane stalled completely enough so that it will spin—unless you use partial power combined with full-back stick.

This is one of those small odd facts concerning an airplane that a good pilot should know; but at the same time it should not obscure the main fact—that the speed of the airplane is controlled by the position in which the pilot holds the stick.

WHAT'S THE POINT?

All this, you may think, is fairly interesting, if true; but it is sort of useless knowledge. What the pilot wants to know is, in the last analysis, how he can make the airplane go where he wants it to go. Speed, Angle of Attack, and all that sort of thing may be very interesting, but what he wants to know is this: what is the effect of the elevators upon the *flight path?* Unfortunately,—this is just the point—no straightforward simple answer is possible. It all depends. A change of speed and Angle of Attack may have one effect in a climb and an altogether different effect in the glide; it may affect the flight path differently at fast speed and at slow speeds. Moreover, any change of stick position has usually two effects on the flight path, first a quick immediate effect (usually a ballooning when you pull the stick back) and then a steady long-run effect (sometimes a downward deflection of the flight path). The elevator's effect upon the flight path must therefore be studied separately for each of the various conditions of flight—the climb, the glide, slow flight, fast flight, high-altitude flight, and so forth. This is done in other chapters of this book.

DESTRUCTIVE CRITICISM

If this discussion of the elevator seems perhaps not very constructive, that is as it should be; it is meant to be positively destructive.

What must be destroyed, what must be completely uprooted from your mind is the idea that pulling the stick back will make you go up. Too many pilots have stalled in or spun in because they tried to use the "elevator" to "elevate" themselves.

We can guess pretty well what happens in the mind of a pilot who stalls in or spins in. The trouble begins dozens or hundreds of hours before the accident actually happens. It begins with a mistaken notion that the stick is an up-and-down control.

This notion is reinforced by the name of the control, the fake erudition that prefers the Latin and misleading word, elevator, to the honest old pilot's word, flippers.

The fatal error is reinforced also by the now fashionable methods of using the controls to maintain level flight. It is true that the "cheapest" way of getting back to your assigned altitude (after an updraft or downdraft has got you off) is to use the stick, rather than the throttle. But so is the cheapest way of making a small course correction simply to kick rudder. As to directional control, we have found out that the cheapest way is not the best from the standpoint of flight training, and a student is today expected to make even the smallest change of direction by a properly banked, properly coordinated turn. Perhaps the method of maintaining level flight which is at present considered correct—to keep the power constant and change altitude by flipper—is wrong from the standpoint of flight training; perhaps we should demand constant air speed and regulation of altitude by throttle—even if that should mean slightly less positive and accurate altitude keeping. It would instill right ideas rather than wrong ideas.

Wrong ideas are cultivated also by our present arrangement of the controls. This arrangement hasn't changed for 30 years, and it seems overdue for a change. Perhaps we shall see airplanes with an up-and-down lever which controls throttle and also spoilers, drag plates, or flaps, in the place of honor; and the stick, atrophied into a glorified stabilizer (trim-tab) control, mounted somewhere at the pilot's side as a speed lever. Who knows?

I WANT TO STAY UP

Be that as it may; with the mistaken "elevator" notion in his mind, the victim some day gets into a situation where he approaches the ground and doesn't want to. (It may be a forced landing, but more likely he is falling out of a steep turn near the ground—it doesn't matter just now. Most fatal accidents actually happen out of turns. And into their making there goes misuse not ᴐnly of the elevator, but of aileron and rudder as well. But for the present analysis it may be best to disregard the rudder and aileron part and concentrate on the pilot's handling of the elevator alone.) Afraid of the ground, the victim tries to set his controls for "up" or at least for "not so much down," very desperately so. Actually, the stick being the speed control, what he does is to put on the brakes! He would know better, if you asked him, than to slow the airplane up in a critical situation. His trouble is not that he doesn't understand the necessity for keeping speed; his trouble is perhaps not even that near the ground he doesn't have the courage to do the right thing—to let the stick go forward. His real trouble is perhaps that someone has switched the labels on his controls for him. He thinks he is pulling the "up" lever. He doesn't know he is pulling the brake—the Angle of Attack lever which if pulled far enough, will necessarily and inevitably stall the airplane!

THE CASE OF THE CRAZY DOG

Because his trouble is due to switched labels, he has no time to catch on. Imagine for a moment that by some strange brain disease your dog got his signals crossed and started to obey your command "Lie down" by running and jumping up on you, and your command "Come here" by lying down. How long it would take you to discover that! You would suspect almost everything rather than the truth. You would think you had not spoken loudly enough, or harshly enough, or that he needed a good whipping. And you would probably get good and mad before you finally caught on. In the same way the pilot who discovers that his "elevator" fails to get the desired results, that is, fails to hold him up, is not likely to suspect that he has done the wrong thing; he is much more likely to think that he has done the right thing, but not hard enough. When his moderate pull on the "elevator" fails to elevate him, he thinks a big pull will. And so he pulls.

The wreckage often shows how desperately hard he pulls. Sometimes the crash fixes the controls in the position in which they were at the moment of impact, and the stick is then usually found hard back. Sometimes the stick is found actually bent, showing how the pilot was hauling back on it in a veritable cramp of desperation, using it as if it were a crowbar with which to pry the airplane's nose up away from the ground.

Epitaph: "He did not know he was pulling the stalling lever."

It is easy to be safe.

THE SILVER CHAIN

The so-called "elevator" is really the airplane's speed control, or, if you want to state it accurately, its Angle of Attack control. This shows how you can absolutely keep yourself from stalling or spinning in!

Stalls and spins are caused by having too low a speed or—what is practically the same thing—too high an Angle of Attack. All right: simply do not move the stick into the position where it will force the airplane to fly at stalling speed (or at stalling Angle of Attack). Simply do not let that hand creep back. For an airplane cannot stall and cannot spin unless the stick is held back in the stalling position! Remember this simple fact. It is of the utmost importance for every pilot; unless the stick is in the stalling position, an airplane cannot stall and hence cannot spin. Stalling position of the stick in most airplanes is nearly all the way back if the power is off, and perhaps about two-thirds back if the power is on. Just what it is in your airplane it is very much your business to find out.

Now from this you might easily get the idea that to keep yourself from stalling and spinning in all you'd have to do would be to buy 10 cents' worth of wire and tie the stick loosely to some structural part of the airplane; so that regardless of how confused or panicky you might become in some emergency, you simply could never pull the stick farther back than so far—say to the position for 15 degrees Angle of Attack or say a speed of 10 m.p.h. above stalling speed. Thus, you might think, the airplane would be rendered unstallable. This would then solve one of the biggest problems of aviation, the terrific death rate from accidents, since most fatal accidents involve a stall. And it would also cut out much tedious training—all the practice of stalls and spins. You might think you had a big idea there!

THE FOOLPROOF AIRPLANE

The strange thing is that you would be right! Ten cents' worth of wire will make any airplane unstallable, will solve one of aviation's biggest problems, and will simplify flight training enormously. "Restriction of controls" is an important new trend in aviation. It is the main principle of all the simplified, "family," or "foolproof" airplanes—a simple, mechanical stop somewhere in the control system that makes it impossible for the pilot to pull the stick back far enough to stall the airplane. The actual engineering of such a control restriction is not quite simple. It has been explained above that in the average airplane a given back position of the stick will bring the air‐ plane nearer the stall when the power is on than it will when the power is off—because the propeller blast hits the control surfaces. Hence a stick restricted so that with power off it could barely not stall the airplane might nevertheless stall it while the power is on. On the other hand, a stick so restricted that even with power on it can't stall the airplane would with power off be so ineffectual that the pilot might not ever be able to slow the airplane up for a reasonable land‐ ing. Hence, the airplane must be so designed that power will not change its "trim." This is done in one of the "foolproof" airplanes by mounting the power plant so that it pulls 10 degrees downward as well as forward.

Another complication: when an airplane is even only near the stall, misuse of rudder or ailerons—without further pull back on the stick—can bring on a stall on one wing, and hence a spin. Thus the designer must make sure rudder and ailerons won't be misused. He must either abolish the rudder altogether, as has been done in some airplanes, or he must connect the rudder mechanically with the aileron control so as to make misuse of the rudder impossible; or he must put some mechanical stop also into the rudder and aileron system.

But that, too, is only a matter of 10 cents' worth of wire. If you tied all three controls down so that none of them could be moved very far, you would still have perfectly sufficient control for all ordinary flying except perhaps three-point landings and steep sideslips; and at the same time you would have a "foolproof" airplane, incapable of spinning. There are several makes of airplanes, built to this formula and "characteristically incapable of spinning," now in use.

One expert pilot says that before he will let his wife fly, he is going to buy her a pretty silver "safety chain" and fasten it to her ship's stick. Actually it may not be a good practice to make changes on your airplane's controls, and it might be of doubtful legality, since a license is issued to a ship "as is." But the best place to put that safety chain is in your own mind. If you can't foolproof your own airplane, you can foolproof your own mind. Just remember that an airplane cannot stall or spin unless the controls are set for stalling and spinning; and keep that hand from creeping back.

Chapter 10

THE AILERONS

WHEN the Wright brothers invented the airplane, the thing they really invented, the key thing, was quite small. It was the device which banks and unbanks the airplane, the device which we now call the aileron. All the other elements of the airplane were then already in existence: men had flown on wings before; motors were quite highly developed; propellers had driven airships on long flights; the rudder was as age-old as the boat; the elevator was nothing but an up-and-down rudder; stabilizing tail surfaces were being used on airships and had been used from time immemorial on arrows; but the aileron was created new and was the one thing that made it possible to combine all those other elements into a flyable machine.

Strictly, in the case of the Wrights, one should not speak of the aileron but of "lateral control." What the Wrights used was not the aileron as we know it now, but the expedient of warping the wings in opposite senses, bending the whole airplane out of shape. Glenn Curtiss used ailerons, that is, movable control surfaces hinged into the trailing edge of the wing tip. A gigantic lawsuit soon developed concerning this point and retarded American aviation for years. Whether or not the hinged ailerons and the warping wing were essentially the same thing was the controversy.

For the following discussion the two are exactly the same thing. Aileron or wing warping—either method works on the same principle: by increasing the Angle of Attack on one wing tip and decreasing the Angle of Attack on the other wing tip, you get more lift on one side of your airplane and less lift on the other side, and your airplane banks or unbanks at your will.

THE ADVERSE YAW EFFECT

All this history may seem a little dull, and the mechanics a little obvious. But look what happened to the Wrights. The idea was sound and apparently quite simple. Yet, when they tried it out for the first time in actual flight and warped the wings of their glider for a bank to the left, what they got was a bank to the right, and a turn to the right, and a crack-up! And fundamentally the same thing can still happen to you today—if you don't watch out. If you ever break your neck in an airplane, your ailerons will probably have much to do with it.

For the aileron has two inherent faults. The more serious of these if the *adverse yaw effect*. When the pilot moves the stick, say, to the right, the ship does respond by banking to the right, as it should; but it wants also to yaw ("turn") to the left—as it should not! The pilot

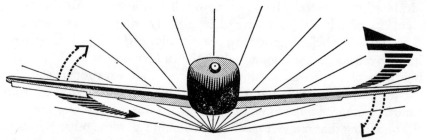

The adverse yaw effect. Ailerons have *two* effects. *White arrows:* rolling effect is what pilot wants. *Black arrows:* yawing effect is an undesired by-product of rolling effect. Here, the pilot banks to (his) left, presumably in order to turn to the left, but the airplane at the same time yaws to (pilot's) *right.* That's why this yaw effect is called "adverse." Pilot must kill this effect by using rudder.

banks the ship to the right probably (not necessarily, but probably) because he wants to fly a curve to the right; but as he banks it, the ship swerves to the left! And this little trick complicates our entire technique of flying—as we shall see.

What causes this behavior? Almost everything about an airplane can be understood by simply taking a good look at it; certainly the adverse yaw effect can be explained to a student without even leaving the ground. Set an airplane's ailerons hard over (stick all the way, say, to the right-hand side) and then stand in front of the airplane and look at it. It then becomes apparent that the downward-deflected aileron (the one that is set to *lift* its wing) projects deeply down into the air stream and will cause, along with much additional lift, much additional drag; while the upward-deflected aileron (the one that is set to *depress* its wing) hardly shows at all, is not exposed to the air stream, and hence has, along with very little lift, very little drag. This is only reasonable and to be expected. You can't have something for nothing. You can't have lift without drag; more lift "costs" more drag. Thus, with the differential *lifts* of the two wings tend to *roll* the airplane to the *right;* but at the same time the unequal *drags* of the two wings tend to *yaw* it to the *left.*

An expert will object here and point out that the ailerons cannot be held wholly responsible for the adverse yaw effect; that some of this adverse yawing tendency is due simply to the rolling motion of the wings and would persist no matter what device might cause the wings to roll. But the explanation just given answers the purposes of the pilot. Even though it does not tell the whole truth, it tells truth, and it has the advantage that it can be "shown."

This adverse yaw effect of the ailerons is a small thing, but one of the really important things to understand about the art of flying. In the first place, it is the real reason why the airplane has a rudder, and why the pilot must work the rudder almost continually. Contrary to the ideas of most student pilots, the rudder is not the airplane's right-and-left steering control; right-and-left steering is done by banking the airplane right and left. The rudder is merely a device by which the pilot counteracts the adverse yaw effects which inevitably arise every time he banks and unbanks his airplane. At the same time, the rudder is without doubt the most difficult of the airplane's controls to master, and the one that uses the most training time. Furthermore, misuse of the rudder is a factor in almost all accidents—you can see that the adverse yaw effect gives rise to a whole string of consequences. All this however—the whole connection between aileron, adverse yaw, rudder, and the airplane's turning—is so important and so intricate that two subsequent chapters will be largely devoted to it. Hence we shall not go into more detail here.

Next, there is the effect which the yawing has upon the lift of each of the two wings. The wrong-way yaw can actually also produce a wrong-way roll! It happens like this: You set your ailerons so as to lift the left wing. This brakes the left wing, and, if the pilot is not alert with his rudder, it makes the ship yaw to the left. Because of this yawing, the left wing tip has temporarily less forward speed and less wind, and hence also less lift; and thus, despite the aileron's being set to raise that wing, the wing may fail to come up! At the same time the yawing gives the right wing tip more speed, more wind, and more lift; and, despite the aileron's being set to depress that wing, the wing may fail to go down or may even go up! This extreme form of adverse yaw is what happened to the Wrights on their first tryout of their ailerons. In a modern airplane with good ailerons and good

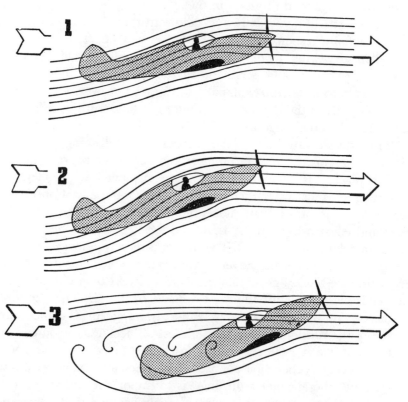

An ordinary stall occurs simply because the pilot forces the airplane *as a whole* to an excessive Angle of Attack. The pilot is trying to lift his airplane by pulling the stick back. He overdoes it, and thus he drops it.

stability, this cannot possibly happen in normal, healthy flight; but it does sometimes happen when an airplane is stalled or nearly stalled and if the ailerons are then used abruptly and to excess—as we shall presently see.

AILERON STALL

Another trick of the ailerons is wing-tip stalling. Under some conditions, an aileron that is set to lift a drooping wing may actually stall that wing and drop it viciously.

To understand how that can happen, a student must first get rid of the mistaken idea—so popular with pilots and so fatal—that a stall is caused simply by lack of speed. He must understand what has been

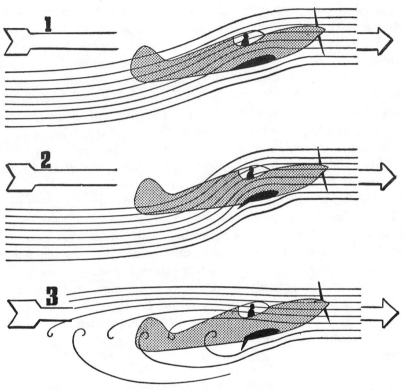

An aileron stall occurs because the pilot forces *one part of one wing* to excessive Angle of Attack. At 1 the airplane is flying at fairly high Angle of Attack to begin with. At 2 the pilot pushes the stick to the left, thus deflecting the right aileron downward and increasing the Angle of Attack of the right wing tip. At 3, he overdoes it, and the right wing tip stalls. The pilot was trying to lift his right wing by pushing the stick to the left. He overdoes, and thus he drops it. The airplane will now enter a spin to the right.

stressed so much earlier in this book—that a stall is caused by excessive Angle of Attack. Whenever a wing meets the air at excessive Angle of Attack, it will stall. And he must once more clearly envisage the nature of the aileron: that the aileron is essentially nothing but a device for changing the Angle of Attack of the wing tip. Only if he understands the nature of the stall and the nature of the aileron can he understand how excessive use of aileron can stall a wing tip.

Here is how the aileron may stall the wing. Suppose that an airplane is flying fairly near the stall—in a slow glide or a steep climb. Both wings are then going through the air at fairly large Angle of Attack, and the pilot realizes that if he held the stick back much farther and set his ship to still higher Angle of Attack, both his wings would stall. What he perhaps does not realize is this: By using the ailerons sharply to raise one wing, he will do to that *one* wing exactly the thing which holding the stick farther back would have done to *both* wings; pulling down the aileron has the effect of setting the whole wing tip to a higher Angle of Attack. And, since that wing is already doing pretty nearly its best, since it is already flying at pretty nearly the highest Angle of Attack of which it is capable, the effect of further increase in its Angle of Attack is that it will stall.

This is what in the older ships used to produce the effect of the aileron's "going out" when the ship approached a stall; the increase in Angle of Attack of the wing tip failed to produce an increase in lift. And this is also what made those older ships so much more dangerous to fly; in extreme cases, the drooping wing not only failed to come up but actually went down, being completely stalled by the very aileron that was meant to lift it! In most modern ships, the "going-out" of the ailerons is no longer so clean-cut, because of the many refinements that have been made on the aileron system.

SOPHISTICATED AILERONS

Because the aileron is the trickiest part of the airplane, much thought has been expended on it, and many improvements have been made. On many airplanes today, those two inherent faults of the aileron have been reduced almost to the vanishing point. In fact, this is one of the few respects in which the airplane of today differs essentially from the airplane of 1918—more trustworthy ailerons. The stall effect especially is no longer of much practical importance; experts say that such aileron troubles as still remain (which to a pilot sometimes look like stall troubles) are actually merely excessive adverse yaw trouble. At any rate it may be interesting to look at some of these improvements.

One important improvement has been made, not by improving the aileron itself, but by improving the *wing* of which the aileron is a part. By various arrangements designers make sure that, as the airplane is slowed up, the stall will progressively spread along the wing from the wing root toward the wing tip, rather than occurring all along the span at the same time and rather than occurring perhaps at the tip first. Thus, when the ship is all slowed up and the center part of the wing span is already stalled, the wing tips still have a reserve to lift which the ailerons can "get." The simplest way to achieve this is "wash-out," a twist of the wing that makes the wing

Wash-out. The wing tip is set at a lower angle of incidence than the wing root. If the airplane's Angle of Attack is increased, the wing root will reach stalling angle first. The stall will thus extend out to the ailerons and wing tips only if the pilot then pulls the stick back even more.

tip ride always at flatter Angle of Attack than the wing root. In many airplanes, this twist in the wing can be seen with the naked eye. Another way of getting the same result is a change of wing *sections*, employing at the wing tip a section that will "hang on" especially well at high Angles of Attack. Still another way of getting the same results is a *slot* in the leading edge of each wing, near the tips—familiar to all pilots from many ships. A slotted airfoil can go to higher Angle of Attack than can an ordinary one; hence a slotted wing tip will remain unstalled when a nonslotted one might be badly stalled; hence, slots in the *leading* edge of the wing tip will make the aileron on the *trailing* edge of that same wing tip much safer and much more effective in slow flight.

Another important improvement has been made, again not by improving the aileron itself, but on the *transmission* between the aileron and the pilot's stick. This is known as the *differential aileron* and is found today on most airplanes. The linkage between the stick and the control surface is such that, when the pilot moves the stick, say, to the right, the aileron on the *left* wing is deflected downward only comparatively little. The aileron on the *right* wing is at the same time deflected upward a great deal. The logic behind this is as follows: Deflecting an aileron upward can't do much harm; it is the downward deflection that causes our troubles—adverse yaw and wing-tip stalling. Hence reducing the downward deflection reduces the troubles, while at the same time the increased upward deflection of the other aileron keeps the whole aileron system fully effective.

The modern aileron is no kitchen door. *A*, Set to depress its wing, the aileron also produces drag, thus minimizing the Adverse Yaw effect. *B*, Set to lift its wing, the aileron produces a slot effect which minimizes stalling tendencies. *X* marks hinge.

Finally, the *control surface* itself has also been improved. The modern aileron has lost all semblance to the original kitchen-door-like simple flap. The *Frise* aileron is based on two ideas. One is the offset hinge. When the aileron is deflected upward, so as to *depress* the wing, this trick hinging causes the leading edge of the aileron to protrude, liplike, down into the air stream. This lip acts as a drag, tending neither to lift the wing nor to depress it, but simply holding it back; and this drag partly balances the drag which at the same moment is being caused by the other aileron, the one which at the same moment is set to raise the other wing. Thus the offset hinge equalizes the drags of the two wing tips and reduces the adverse yaw effect.

The other idea behind the Frise aileron is the slot effect, which goes into action when the aileron is deflected downward, so as to lift "its" wing. This slot has its usual effect of preventing a stall; and thus the Frise aileron remains comparatively effective at very high Angle of Attack when a simple kitchen-door aileron might stall the wing instead of lifting it.

Despite all these improvements, both those inherent faults of the aileron, adverse yaw and wing-tip stalling, are still present in almost all our airplanes and are likely to cause trouble in certain situations when you least expect them—especially the adverse yaw. It might be a good idea for the student to convince himself experimentally that those effects are still present, that they are not theoretical abstractions but actual and important facts.

One way to do this is to try some stalls and see how the ailerons really work. This is a valuable experiment. When an airplane is stalled, it becomes laterally unstable, that is, it wants to drop off over one wing or the other. The natural reaction of the pilot is then to try to keep it level by using his ailerons. But in stall practice, the student is told that this is wrong; he must never use the ailerons while stalled. Instead he must maintain lateral control by quick decisive footwork on the rudder. He gathers that this is because the ailerons might stall the wing badly and cause a spin. Sooner or later, though, he is bound to experiment while flying solo and use aileron just the same. And he will find in most ships that the ailerons are quite nicely effective even in the stall—that all this talk about wing-tip stalling is not really true—at least not in his ship. This, however, is a dangerous sort of assurance. For the ailerons are effective only because he is using a good deal of rudder at the same time; for, at this stage of his training, a certain degree of control coordination has become second nature to him.

If you really want to find out what your ailerons do to your airplane in a stall, take your feet off the rudder, stall the ship, and, as the ship fixes to fall off, say, over the left wing, try to keep it under control by ailerons *alone*. What you will probably observe is that your ailerons make a brave attempt to raise the drooping left wing and seem for a while to be succeeding; but at the same time the adverse yaw effect—now unopposed by any rudder—slues you around to the left so badly that the wing never succeeds in actually coming up; the ship will probably yaw clear around by 180 degrees and will eventually spin off over the left wing, although you are trying to hold it up with the aileron, or really *because* you are trying to hold it up.

It is interesting also to do a spin and at one point suddenly put the aileron against the spin—the thing which the naïve, untutored flier would naturally do if he found himself suddenly in a spin. Most airplanes will immediately begin to spin much faster and more viciously, owing to adverse yaw or wing-tip stalling or both. It is not advisable to try this without the foreknowledge of an instructor, since in some ships it might lead to some dangerous type of spin.

Still another occasion when the student can observe the adverse yaw is in taxiing against the wind; the ship will tend to turn right if stick is held to the left, and vice versa. This is especially noticeable when taxiing a seaplane and is an important trick in seaplane handling.

A demonstration of the adverse yaw effect in ordinary cruising flight, simply by rolling the ship from side to side, may remain somewhat unconvincing to a student, because in modern ships this effect has been so well compensated for that in fast flight it is almost negligible. But try the same thing with the airplane in fairly slow flight (stabilizer wound well back), and you will get the effect much more sharply.

HOW TO FLY WITH KITCHEN DOORS

In short, then, even the present standard airplane must be flown somewhat as if it had the kitchen-door type of aileron—only that the modern pilot may be a bit sloppier about it. For the sake of clearness, it may be best to assume for a moment that all these improvements had not been made; and to think over just how a pilot would have to behave in an airplane that had just simple kitchen-door ailerons.

Certainly he would find out right away that he must never use the ailerons without diligently supporting them by rudder. This would be a valuable discovery for a student to make, for in fact this is the main purpose of the rudder in all airplanes, old and new: to give the pilot a means by which to combat the adverse yaw effect of the ailerons. That is what "coordination" is all about; that is why we use rudder on going into a turn and coming out of a turn; if the ailerons had no adverse yaw, we would not need a rudder.

And what a blessing it would be for the beginning student if he could early in his flying career get a clear concept of what he has got a rudder for! How much skidding and slipping it would eliminate! How many necks it would save! Too many fliers still stagger around with the idea that the purpose of the rudder is to steer the airplane 'round the corner; this notion is about as healthy as that a rattlesnake makes a pleasant pet.

Next, the pilot of the kitchen-door airplane would discover that his rudder action must become more and more lively the more he slowed the airplane up. Even while he might be still far from stalled he would have the feeling that he was maintaining lateral control mostly by his footwork—the way you maintain it in your present (non-kitchen-door) airplane in a real stall. An example of this would be the climb out of a small field, perhaps on a gusty day. Old-timers may still remember how this used to be a thing which a good instructor would stress, one of those priceless bits of wisdom that one didn't get from routine instruction and that were definitely not in the books. "Now on a day like this," he might say, "don't rely on your ailerons too much. If the left wing drops on you, sort of kick it up by right rudder." To a more recently trained pilot, this might sound as if in those ships you must have been always close to a stall in a climb; but of course you were not. It was merely that the ailerons had more adverse yaw effect; they also were less effective, and you had to come farther over with the stick to get results, which again increased the adverse yaw effect. So that, without lively rudder action on the pilot's part, the ship would have responded to the ailerons simply by yawing: the left wing, instead of coming up, would have stayed down; and the drag of the left aileron would have nosed the ship sharply around to the left.

In a modern airplane, this effect is very much washed out; but it does exist. The more you have your ship slowed up, the more footwork on the rudder you will need to "coordinate" with your handwork on aileron.

HOW TO GET OUT OF A TURN IN A HURRY

Flying our imaginary kitchen-door airplane, a pilot would make an interesting observation concerning the recovery from a steep turn. He would find that, banking into the turn, he could manage with very little rudder. But on trying to unbank, to come out of a steep turn, he might find it impossible to bring the wing up without much help from the rudder; the adverse yaw effect would keep pulling him into the turn; so that with his feet off the rudder he might be simply helpless and might well wind up in a spiral dive or a spin. Why is this?

In a steep turn the airplane flies at high Angle of Attack—the pilot holds a lot of back pressure on the stick. The higher the Angle of Attack of the airplane, the more pronounced is the adverse yaw effect. Banking out of fast, low Angle of Attack flight into the turn, there is not much yaw and hence little need for rudder. Unbanking out of slow, high Angle of Attack flight, there is much yaw and hence much need for rudder.

Right here, you probably think that all this is too theoretical and of no practical value. But actually this very effect has broken many a neck. Most fatal accidents develop out of steep turns near the ground. There is reason to suspect that in many cases the pilot stumbles over his ailerons, so to speak. Turning and steeply banked, with the stick well back and a lot of g load on the ship, he decides that something is wrong. Just what that something might be makes another story; it might be an incipient stall, or a gust that suddenly overbanks him, or what not. At any rate, he suddenly wants to get back to straight and level flight in a hurry. He puts on extreme aileron to the high side, so as to unbank. But because he is then flying at high Angle of Attack (with the stick well back) the adverse yaw effect is very powerful and "pulls him in," swerving his nose to the inside of the turn, and downward. To this swerving of the nose the panicky pilot is likely to react by still more high aileron (to get unbanked) and still more back stick (to get the nose up), and thus he works himself into a spin; but what actually happened was that he stumbled over his aileron.

In our imaginary kitchen-door airplane, there would be only one way to get out of a steep turn quickly and safely—especially once that steep turn had begun to go sour. *First* let the stick come forward, thus reducing Angle of Attack, getting the wings unstalled, giving the ailerons less adverse yaw, and also making sure that the ailerons won't stall the wing rather than lifting it. *Then* a lot of "top" rudder (left rudder if in a right turn). And *then* only high aileron, so as to roll the airplane out of the bank. Simple ailerons used on a slow-flying or nearly stalled airplane (and any airplane is in that condition in a steep turn) would be useless or even nasty. The other controls would first have to create the proper conditions before such ailerons could become effective or even safe to use.

Here again, what is true of the kitchen-door airplane is true, in a washed-out fashion, also of more modern airplanes. Even the modern airplane needs more rudder coming out of a steep turn than going into one. Even in modern airplanes, if a steep turn goes sour, there is only one safe way to recover in a hurry—get the stick forward first, put on top rudder then, and last of all, and *gently*, use your ailerons. And if such a recovery produces a certain amount of sideslip or other "lack of coordination," never mind.

Stick forward! That would quite in general be the main thing a pilot would learn in flying an airplane with kitchen-door ailerons. Whenever the ship would be slow to respond to the ailerons, he would learn not to put on more aileron, holding the stick farther over, but instead to get the stick forward and thus revitalize his ailerons. He would learn this, or else he would sooner or later spin in, pulled into the spin by excessive adverse yaw, or dropped into the spin by a wing tip stalled through excessively used aileron. And the same general rule is still true of today's airplanes, and very much of gliders. Whether you are trying to bank or to unbank or whether it is a gust that hits you: *whenever* the airplane fails to respond readily to a little aileron, the thing to do is not to use more aileron but to get the stick forward. Sometimes, it is true, a lot of aileron will do what a little aileron failed to do; usually, it is true, energetic help from the rudder can cope with a situation that aileron alone can't handle; but the only certain way to regain control surely and promptly, to maintain control under all conditions, is to get the stick forward.

Chapter 11

THE RUDDER

THE important thing to understand about the rudder pedals is that they are unnecessary; like your wisdom teeth, they serve no very good purpose but can cause much trouble. The airplane needs no rudder pedals. It should have no rudder pedals. In all probability it will have no rudder pedals 10 years hence.

Yet, as long as our airplanes have this control, it is most important to the pilot. It is the control that causes the greatest difficulty for beginners. Fully a third of an elementary course in flying is devoted (when you come right down to it) to teaching the use of the rudder, its proper "coordination" with the other controls. And even the more experienced pilot often has trouble using it correctly. In the typical fatal accident, which involves a stall and a spin, misuse of the rudder is almost always partly to blame—along with two other pilot errors already discussed, an elevator pulled back too far in an attempt to keep the airplane up, and a possible "stumble" over the ailerons.

As you might expect, there is a connection between the fact that the rudder control ought not to be on the airplane at all and the fact that pilots have so much trouble using it correctly. You would have trouble using a stick of dynamite correctly in your children's nursery.

THE USES OF THE RUDDER

To clear this up, let's go down the list of the various purposes for which the rudder is used in our present conventional airplanes.

First of all, to steer the airplane while *taxiing:* It is in taxiing (and *only* in taxiing) that the rudder pedals of the airplane get results similar to those of a car's steering wheel. But at best, steering the airplane by rudder is a nerve-wracking job; for at the slow speed of taxiing, the air doesn't flow past the rudder briskly enough to make it very effective. Hence, almost all airplanes are now equipped also with other steering devices for use on the ground. Independent wheel brakes allow the pilot to brake one or the other wheel and thus steer

the airplane. In taxiing a twin-engined ship, differential use of the
two throttles turns the airplane right or left. Many training ships have
steerable tail wheels linked up with the rudder pedal. Of those which
have tricycle landing gears, some have a steerable nose wheel. Thus
the rudder's services in taxiing do not make it indispensable.

The rudder is fairly important during the *take-off run*. When the
steerable tail wheel (or the steerable front wheel, as the case may be)
is no longer on the ground, the airplane can't be steered by it. Steering
by brakes, on the other hand, would be clumsy during a take-off run,
when you want all the acceleration you can get; and thus the rudder
is vital during the take-off run. However, tricycle-geared airplanes
have been built which keep all three wheels on the ground until the
very moment when the airplane breaks away from the ground and
becomes controllable by the usual flight controls. Such an airplane is
therefore entirely controllable, while on the ground, by steering the
nose wheel; it needs no rudder. In such airplanes, the nose wheel is
linked up mechanically with the same control that moves the ailerons,
and the rudder pedals get a fate they deserve: there aren't any!

In flight, most airplanes need rudder against the "torque," that
annoying tendency of an airplane to turn to the left under the influence
of its power plant. If an airplane does suffer from torque, it does need
a rudder—and all single-motored conventional airplanes have this
trouble at least in some flight conditions. But here again, airplanes
have been built that are free of this disturbance, proving it can be
done; hence this use of the rudder is not essential either.

Sometimes the rudder is used to produce *sideslip*. If you put on
left aileron, the airplane will bank to the left; once banked to the left,
it will want to turn to the left. But if you put on right rudder along
with the left aileron, the airplane cannot turn to the left but will
instead slide sideways—downward to the left. Sideslip is useful in
maneuvering an airplane to a landing on a predetermined spot, since
it permits the pilot to get rid of altitude without picking up speed.
But the maneuver is unnecessary in an airplane equipped with a
flap, a spoiler, or some other device by which the descent can
be steepened without building up excess speed. Not all airplanes
have such a device, but all should have one, and soon probably
all will have one. Hence, this use of the rudder, too, is not really
essential.

The rudder is also used in *cross-wind landings*. In order to have the airplane going straight over the ground at the moment when it makes contact with the ground, it is made to sideslip through the air, by the same disposition of stick and rudder that has just been described. For example, if the cross wind is from the left, the stick is held to the left, rudder to the right. This maneuver is admittedly essential in a conventional airplane. But an airplane equipped with tricycle landing gear need not go so strictly straight over the ground at the moment of ground contact; it can afford to touch with some sideways drift. Hence, this use of the rudder could be dispensed with, too.

The rudder is very important in a *stall*. When stalled, the airplane becomes "laterally unstable," that is, wants to capsize to the left or to the right. Ailerons don't work well on a stalled wing; as explained elsewhere, they are likely to drop the wing they are intended to lift and to promote a spin. What the pilot really ought to do, of course, is to get the stick forward and let the airplane come out of the stall. But as long as it is stalled it is true that only the rudder will keep its wings level or keep it from spinning. And, once the airplane is in a spin, it is true that the rudder is an important help in getting it out.

This is the main reason why the present-day airplane still has a rudder. The catch is, however, that there is no good reason why an airplane should ever be stalled, let alone why it should be allowed to spin. The spin has no utility whatever, not even as a combat maneuver of evasion. Airplanes can be made unstallable, and they can very successfully be made unspinnable by restricting the elevator travel and suitably designing the ailerons. And once the stall and spin danger is gone, this use of the rudder is also no longer essential.

And so—the rudder has no real function in flying.

BUT WHAT ABOUT THE TURN?

"But wait a minute," says the student pilot, "What about turns? How could you ever turn the airplane if you had no rudder?" And this is where we get down to business, important business: to analyze how an airplane really turns and what exactly the rudder does in a turn.

Your kid brother thinks that an airplane turns because the rudder is held over; he thinks that an airplane's rudder pedals do in the airplane exactly what the steering wheel does in an automobile. Almost every beginning student has the same idea; and even quite a few experienced pilots still will tell you the same, though they don't actually fly that way. It's still in many books—what's worse, it is still being taught to kids in high schools! That idea is the worst single misconception about flying that is today still widespread. It is a curse. It causes more headache to student pilots than any other single thing in flying; almost every student wastes a lot of time because he can't figure out what to do with the rudder and when and why. It also kills more pilots than any other cause, save perhaps that other idiocy, the idea that pulling the stick back makes the airplane go up. Or rather, it is the combination of the turn-by-rudder fallacy with the up-by-elevator fallacy that is sure poison.

The airplane does *not* turn because the rudder is held over; it turns because it is banked. A turn is flown by leaning the airplane into a bank and then lifting it around, as it were, by back pressure on the stick. The rudder has no primary role in the whole procedure. This may be a new and strange thought to you; it is so overwhelmingly important to anybody who trusts his neck to an airplane that a special chapter will be devoted to it, trying to give you a clear idea of how an airplane really does turn. Meanwhile, you may take it on faith. It is not, as you might possibly think, a pet theory of the writer's. It is accepted by all good instructors and all designers. One definite proof of it is that at least two makes of airplanes have been built which have no rudder at all—not only no rudder pedals, but actually no rudder—and they are nevertheless completely maneuverable, able to execute turns with normal quickness.

"It still does not make sense to me," a student may say. "You claim that turning flight results not from using the rudder, but from the bank, combined with back pressure on the stick. But at the same time, my instructor tells me that I must use rudder in every turn. If I am not using the rudder to turn the airplane, then what in the name of common sense am I using it for?"

You can't blame him for being puzzled. In every other case where he uses the rudder, he understands clearly what he is doing with his feet; what the force does, so to speak, which he is exerting with his foot. In taxiing, he knows it kicks the tail around to turn the airplane. In take-off and landing runs, he knows it holds the airplane straight against cross winds, "torque," and uneven ground, which are trying to make it swerve. In take-offs and climbs he knows it counteracts the "torque." In a sideslip he knows it counteracts the ship's tendency to turn in the direction in which it is banked. But when he puts foot pressure on a rudder pedal during a turn, he does not know what this

Aileron without rudder: The pilot banks the ship to the left, presumably intending to turn to the left. He uses no rudder. The ship banks to the left but yaws to the *right.* "The ball" rolls off to the left, the pilot feels thrown against the left side of the cockpit. A left turn will eventually result, owing to the bank; but it will have been entered in a sloppy and slipping manner.

force is supposed to do. It is not obvious, and too often the instructor fails to show him.

The student needs to be shown a force that disturbs the airplane during a turn and must be counteracted by rudder. This force is the adverse yaw effect of the ailerons. It is against the adverse yaw effect that we use rudder in a turn—against the adverse yaw effect and nothing else! The moment the student becomes conscious of the adverse yaw effect he can see clearly why he must use rudder in a turn and when and how much.

The adverse yaw effect has been discussed. It is that tendency of the ailerons to pull the airplane's nose around to the left side while they are putting it into a right bank, and to the right side while bank-

Aileron with rudder. As he uses aileron, the pilot also uses rudder. This kills the adverse yaw effect. The ball stays in the center. The airplane simply banks. Once banked, it begins to turn to the left simply because it is banked. The pilot has not used the rudder to turn the airplane; he has used it merely to keep the airplane from turning the wrong way.

ing the ship to the left: the wrong-way turning tendency. This tendency is found in all conventional airplanes; but in many types it has been so successfully reduced, at least for normal fairly fast flight, that an unobservant pilot might fly for years and never become clearly conscious of it. The causes of the effect have been fully discussed in the previous chapter. Our job here is to figure out just how it disturbs the airplane in the turn.

THE ENTRY INTO THE TURN

Suppose, then, that a pilot flies a turn (say, to the right) with his feet off the rudder pedals. As he presses the stick to the right, the airplane leans into a bank to the right. As soon as it is banked to the right, it naturally wants to turn to the right. A perfect turn would thus naturally result—if it were not for the adverse yaw. But in using his ailerons to lean the airplane into the bank to the *right*, the pilot produced an adverse yawing force that wants to swerve the airplane's nose to the *left!* This force now keeps the airplane from going into the right turn as it should and as it naturally would. And thus the airplane is banked but cannot turn; instead, it sideslips to the right.

Just how strong this effect is depends upon the type of airplane, particularly on the type of ailerons it has, and also on the Angle of Attack at which the airplane is flying. It also depends on how hard the pilot shoved the stick to the right. Some ships, especially older ones with unsophisticated ailerons, when flown without rudder, will not turn at all but simply bank and sideslip to the right as long as the stick is held over. Other ships, especially newer ones, have such perfectly designed ailerons that the whole effect is negligible at least in normal fairly fast flight. Most airplanes will actually turn, but will turn hesitatingly and will sideslip a little even while they turn.

Every pilot knows this characteristic hesitation of the airplane at the beginning of a turn. And probably every pilot can remember how in his first few hours the instructor would complain that he was slipping and would tell him to use more rudder.

For that is what the rudder is used for in a turn; it is used to counteract the adverse yaw effect. It rids the airplane of that disturbing influence which keeps it from turning into the bank. When you use rudder in the beginning of a turn, you use it not to turn the airplane, not to "help get the turn started"; you use it only to keep the airplane from turning—the wrong way! The rudder, then, is not of equal rank with the ailerons; it is the servant of the ailerons. You use rudder *because you are using the ailerons*.

THE TURN ITSELF

But now, back to the pilot who is flying a turn without rudder. Suppose that he disregards the hesitation and the slipping and simply establishes the bank he wants—say 45 degrees. The moment he has established the bank, he no longer needs any ailerons, and thus he lets the stick come back to neutral.* And the moment he does that he rids the airplane of all adverse yaw effects. And now the airplane is free to do what it "wants" to do, that is, to turn into the bank. Freed of the adverse yaw effect, it suddenly starts to turn willingly and flies a perfect turn!

Every pilot knows that characteristic change in the airplane's manner—how at one point, when you bring the stick back to neutral, it suddenly seems to become willing to turn. And *now* what about the rudder? Now that no adverse yaw effect acts on the airplane, the rudder is no longer needed. Thus, when he neutralizes his ailerons, the pilot must also neutralize his rudder. If he did not, if he kept holding rudder against a force that is no longer there he would merely disturb the airplane.

This is a very frequent fault with the students; for if you have the usual idea that the rudder is the airplane's turning control—why, naturally you will want to hold rudder as long as you want to keep turning. Probably every pilot can remember how in his first few hours the instructor used to yell during this part of the turn. "You are skidding! You are holding too much rudder!" or, "Stop riding the rudder!"

This would puzzle him greatly. Just how much rudder did it take to fly a turn? First, in the entry to the turn, this fellow had told him he was not using enough rudder. Now, a few moments later, it was too much rudder. You can see now what his trouble was; he did not understand the purpose of the rudder. Had he understood in the beginning that only purpose of the rudder in a turn is to counteract the ailerons' adverse yaw effect, it would have been obvious to him that, in the entry, when the ailerons are working hard, a lot of rudder is needed; while in the turn itself, when the ailerons are doing practically nothing, no rudder is needed.

STRAIGHTENING OUT

But now again, back to the pilot who is flying a turn without rudder. Suppose now that he has turned far enough and wants to

* Important fine points regarding the handling of stick and rudder in a turn are disregarded here, in order to let the main idea appear more clearly.

straighten out. Being in a bank to the right, he now presses his stick to the left, so as to unbank the airplane. What happens? The adverse yaw effect reappears, but on the other side; now that the pilot has the stick to the *left*, the adverse yaw pulls the nose to the *right*. Now that the pilot wants to stop the airplane from turning to the right, the adverse yaw turns it even more to the right!

Just how pronounced this effect is depends again on the type of the airplane and particularly on the type of ailerons it has; and again it also depends on just how the pilot handles the stick. There are airplanes that, roughly handled, cannot be brought out of a turn by ailerons alone. There are others in which this effect is very slight. Most airplanes, flown without rudder, will unbank and will eventually stop turning but will do it unwillingly and sluggishly, with much wallowing, skidding, and slipping.

Every pilot knows this characteristic of the airplane; the moment you press the stick even the least bit to the high side, giving aileron as if to unbank, the nose begins to hurry around along the horizon with a sudden new eagerness. And it is against this eagerness to turn that we use rudder while coming out of a turn. We don't use it, strictly speaking, to "stop" the turn or even to "help stop the turn"; the turn is stopped by getting the airplane out of the bank. We use it to kill the *additional* turning tendency which is suddenly caused by our use of the ailerons; again, the rudder is entirely subservient to the aileron and is used *because we use ailerons.*

So much for the use of the rudder in the turn; it is used, then, purely to keep the adverse yaw effect from swerving the airplane. The whole procedure of flying a turn is described in more detail elsewhere. Right here, another thing must be made very clear concerning the rudder. The rudder must keep working against the adverse yaw effect not only at the beginning and the end of turns, but all the time.

COORDINATION

For we use the ailerons not only when we want to turn and when we want to stop a turn, but we use them almost continually even in straight flight—unless the air is perfectly smooth. Every time a gust drops one wing or the other, we level the airplane again by aileron. And, whenever the ailerons are used, however slightly, they cause an adverse yaw effect; whenever there is an adverse yaw effect, the airplane is not flying perfectly but skidding or slipping slightly; hence, for perfect flying, even the slightest hand pressure on the ailerons must be accompanied by foot pressure on the pedals. *"Both together"* is the rule of smooth flying.

You should convince yourself that the adverse yaw actually does exist and that it actually does disturb the airplane noticeably. One way to do this is to fly the airplane in fairly rough air without rudder; the nose will continually yaw. And if you sight out along one wing watching the wing tip's apparent motion relative to the ground—you will see the yawing even more clearly; how now the wing tip seems to stand still for a moment and then again takes a sudden jump forward —instead of smoothly sliding along as it should. The moment you put your feet to work on the pedals and move stick and rudder "both together," this continual back and forth yawing stops.

The adverse yaw effect can be clearly observed also when taxiing against a fairly strong wind. It is possible to steer the airplane on the ground by using the adverse yaw. If you want to turn left, hold the stick over to the right; now, on the ground, with the ship resting on its two wheels or two pontons, the ailerons cannot bank the ship; but this adverse yaw effect is free to turn the airplane the "wrong" way, that is, to the left. This is an important trick in handling a seaplane on the water.

THE AUTOMATIC RUDDER

The Wright brothers knew what the rudder is for. They had the whole airplane figured out much more brilliantly even than most people realize even now. They knew that an airplane could not be successfully turned by rudder but would have to be turned by leaning it into a bank and lifting it around with the flippers. Their first glider didn't have a rudder! But they also soon discovered the adverse yaw effect: when their first attempt to bank to the right produced a turn to the left, and a crash. The Wrights then fitted a rudder; but they understood the nature of the rudder better than most airmen have understood it since. They knew that it was merely a device for counteracting the adverse yaw effect. They hitched their rudder up mechanically with their aileron control. Thus giving aileron to the right would automatically always be accompanied by right rudder;

aileron to the left, by left rudder; aileron in neutral, neutral rudder. What we now call "coordination of stick and rudder," what we spend tedious hours learning, and—as the accident record shows—never learn quite well enough, was reduced to a mechanical device!

The latest trend is back toward the same idea. In the modern safety airplane, pioneered most successfully by Fred E. Weick, the rudder is once more hitched up mechanically with the same steering wheel that also moves the ailerons—and once more there are no rudder pedals! Such an airplane has restricted flippers and thus cannot be badly stalled; hence it does not need an independent rudder for control in stalls and spins; it has no "torque" and hence needs no independent rudder action merely to keep straight. It has a tricycle landing gear with steerable front wheels and hence needs no rudder for control on the ground. Because of its tricycle landing gear, it can afford to touch the ground in a cross-wind landing with some sideways drift; hence it doesn't need an independent rudder to produce sideslip. Whenever the pilot uses the ailerons in such an airplane, the rudder goes over at the same time. Whenever the pilot neutralizes his ailerons, the rudder, too, returns to neutral. Such a ship is built, then, on the clear-cut theory that the rudder's only purpose is to counteract the adverse yaw effect of the ailerons. And such an airplane flies a perfectly coordinated turn and "coordinates" perfectly also in straight flight through rough air. In very fast flight, when the ailerons' adverse yaw effect is very slight, it may skid a little from slightly too much rudder action; in very slow flight, when aileron yaw is pronounced, it may yaw a little from insufficient rudder action. But experiment has proved that even the most expert pilot cannot in the long run "coordinate" as well as does such an airplane, and that he will occasionally produce skids or slips much worse than such a "two-control" airplane could possibly produce; and of course such an airplane's control action is infinitely better than that of the mediocre pilot or the scared pilot or the confused and tired pilot.

That proves it then: The only real purpose of the rudder is to counteract the adverse yaw effect of the ailerons. And—some designers go even farther. Ailerons *can* be designed that cause practically no adverse yawing effect; hence we ought to be able to do not only without the rudder *pedals* but actually without the rudder itself: without any movable vertical surface on the tail. Some slight yawing tendencies might remain, but they can easily be stopped by the fixed vertical tail fin—if the fin is only big enough and the tail long enough. Along this formula Prof. Koppen of M.I.T. has built rudderless safety airplanes which "coordinate" extremely well; and he goes so far as to say this about the rudder, "The only purpose of the rudder is to cover up the mistakes of the designer."

PART IV

THE BASIC MANEUVERS

There are only four maneuvers that a pilot can execute in an airplane: the turn, straight and level flight, the glide, and the climb. All other maneuvers, however difficult, however intricate, however important they may be, are only variations or combinations of those four fundamentals.

This is not an idle theoretical notion. It is the official doctrine that governs the entire curriculum of modern flight instruction. For example, if the student pilot has difficulties with his landings, he is today not allowed to practice landings until they finally click—as he wants to, and as used to be customary in flight instruction only a few years ago. Instead he is returned to the practice of fundamentals—in this case, to the practice of glides and of stalls, which can perhaps be called a form of the glide. For flight instructors find that, if the student's mastery of the fundamentals of flight is firm enough, his landings will click almost at first try and will require amazingly little practice. Quite in general, if the student has difficulties with any maneuver, the modern flight instructor always suspects that the trouble is some flaw in the student's understanding of the fundamentals.

This may throw some light also on the true role of the so-called "advanced" maneuvers, the chandelle, the lazy eight, the pylon eight, and so forth. To the student at flying school, those maneuvers sometimes appear as the essence of all flying; but the instructor knows that they are not at all important in themselves but are merely exercises by which to develop better mastery of those four fundamentals; and tests by which to show up flaws in the student's technique of turning, climbing, gliding, and so on—flaws that might otherwise go undetected and cause trouble in some practical flight situation. The final aim of, say, the chandelle (that tricky combination of climbs and turns) is still merely to enable the student to fly a good turn and a good climb. Hence this book will not discuss the chandelle, the lazy eight, the pylon eight, but will discuss instead directly the things that those maneuvers try to teach, that is, those four fundamentals of flight.

One might perhaps take exception to the idea of four fundamentals. For example, one of the following chapters will show that straight and level flight is really nothing but a series of S turns, infinitely shallowed out. One might maintain that there are fundamentally only two things to be learned in controlling an airplane—how to fly a turn and how to fly at various Angles of Attack, which includes the stall and the spin. In order to make this book more useful to the reader, however, the official doctrine of the "four fundamentals" is accepted here, and the following chapters will discuss three of those fundamentals—the turn, straight and level flight at cruising speed, and the glide. About the climb, some important facts will be found in Chapter 19 on The Working Speeds of an Airplane. About those phases of flight which involve high Angle of Attack—"mushing" flight, the stall, the spin—some explanation has been attempted in the first chapters of this book.

Chapter 12

THE TURN

HOW do pilot and airplane behave during a turn?

The record shows that they behave badly. A close analysis of accidents (elsewhere in this book) suggests that we pilots, as a group, simply don't know how to turn right or left. Almost all fatal flying accidents are caused by loss of control during a turn! And this despite the fact that out of a student pilot's first hundred hours of flight instruction, at least 50 hours are devoted to teaching him to to make turns—for most of the so-called "advanced" training maneuvers such as chandelles, lazy eights, pylon eights are really only exercises in turning right and left. That may seem strange, even unbelievable; imagine a motorist not really knowing how to turn an automobile right or left; imagine a farmer getting mixed up on the "gee" and "haw"!

But flying is still quite an art.

Suppose you start out by asking your kid brother how a turn is flown. He has read quite a bit about airplanes; he has hung around at the airport. He has taken a high-school course in aeronautics. He ought to know.

THE TURN-BY-RUDDER IDEA

Here is what he will tell you: The control that turns the airplane is the rudder; that's what an airplane has a rudder for, stupid! Except that it is worked by pedals, an airplane's rudder is used exactly like the steering gear of an automobile. When you want to turn left, you press the left pedal forward with your foot, at the same time relaxing your right foot and letting the right pedal come back. The more sharply you want to turn, the farther you hold the left pedal forward with your foot; as long as you want to keep turning, you keep holding the pedals that way; and, when you finally want to fly straight again, you let your two pedals come back to neutral again, and the airplane's flight path straightens out.

That is not, however, your kid brother's whole story. He will next explain to you about the bank. Luckily, he will say, the airplane (not running stodgily with four wheels on a flat road) can be banked as it goes around the turn—the way a bicycle or motorcycle is banked; and is it fun! If you make a sharp turn, you sometimes are 'way over on your side. Boy, it is keen; and although it looks dangerous, it is nothing to worry about. In fact, because the airplane is so fast, and the air is so thin, you have to bank the airplane in every turn, or else it will simply slither on in the old direction even though you turn its nose into some new direction, much like a car that turns too sharply on a slippery road.

And he can tell you just how to bank the airplane: Hold the stick to the side toward which you want to bank. The more steeply you want to bank, he says, the farther over you must hold the stick; and you must keep holding the stick over as long as you want to stay banked. When you bring the stick back to the middle again, the airplane will level its wings.

And finally, he can even tell you just how to "coordinate" your stick and rudder. If you feel a skid, he says, it is a sign that you are using too much rudder and not enough aileron; you are getting too much turn with not enough bank, and you need more aileron. If you feel a slip, he says, it is a sign that you are using too much aileron and too little rudder and are banked too steeply for the amount of turning you do, and you need more rudder.

This now makes a pretty good theory. It is simple, straightforward, and it sticks with you. But it is no good. It causes many men to wash out from Army and Navy flight training—for if you follow it out, you will never coordinate your controls quite correctly. It also causes many accidents; for if you follow it out, you will often fly turns, especially sudden ones, with the controls all set for a spin. And if you get your controls set for a spin often enough, it is only a question of time before some day you do spin.

Here is what's wrong with the theory: It is built on a mistaken idea that the airplane turns because of the rudder. Most of the troubles, small, big, or fatal, that pilots have with turns can be traced to this one idea, that the rudder is the airplane's turning control, that a turn in an airplane can be produced only by putting the rudder over.

Unfortunately, this turn-by-rudder idea is hard to kill. It has a way of re-creating itself. It is often used to explain the airplane's controls to high-school kids or general magazine readers—simply because it is easy to understand for nonflying writers and readers, teachers and pupils. It re-creates itself also in the pilot's own nervous system, even after he really should know better. This is because you can use the rudder with apparent success to "steer" the airplane, that is, to make small changes of direction, in straight wings-level flight. As explained elsewhere, this is a faulty technique of flying straight;

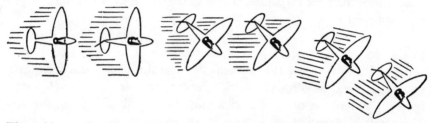

The rudder produces yaw, not turn. At station 2, the pilot kicks the rudder over: at station 4 he is still moving in the old direction! This sideways skidding flight is one of the two meanings of the word "yaw." Even at stations 5 and 6, change of direction but it is sluggish and skidding. *Not shown:* as the airplane yaws, its left wing comes up, right wing goes down; once it is banked, the rudder yaws the nose earthward: Thus the real net result of holding rudder will be, not even the sluggish, skidding turn shown here, but a corkscrew dive! The rudder is *not*, as often stated, the airplane's directional control. *The airplane has no directional control.* Direction of flight is changed by a complicated maneuver, employing several controls in a definite time sequence. This delicate coaxing-around of the airplane is called the turn, and is about half of the art of flying.

but it is widely practiced, and if you use your rudder 30 times a minute to steer the airplane a little bit more to the right and a little bit more to the left, you are bound to use it also when you want to steer the airplane a whole lot to the right and a whole lot to the left!

But the rudder can *never* produce a turn. It cannot "start" a turn or even "help the turn get started." It cannot "stop" a turn or even "help the turn get stopped." The only effect the rudder can ever produce is *yaw*, and the only effect it can ever stop is *yaw*. Yaw, in this sense, means practically the same as "skid" or "slip." The airplane's nose is swung to one side or another while the flight path continues substantially unchanged, so that the airplane slices through the air slightly sideways. Yaw is not turn. A turn, a clean nice curving of the flight path without skid or slip, cannot be produced by the rudder but is produced by entirely different means. The rudder is essentially quite unnecessary for the turn. Some airplanes haven't even got a rudder, but only a rigid vertical fin, and yet they turn. And the birds don't even have a vertical fin!

All this will be made clear to a patient reader. But first we must kill the turn-by-rudder idea. Only when that is done can the reader's mind be really receptive to the story of how the airplane really does turn.

WHAT HAPPENS IF YOU TRY TO TURN BY RUDDER

There is a close connection between saintliness and sin; it is always worth while to find out by actual experiment that sin does not pay, and just why not. Suppose you find out in actual flight just what happens when you attempt to turn the airplane by rudder alone.

Out of straight flight, with wings level, put on gentle right rudder, leaving your hands entirely off the stick. At first it works pretty well. You get a turn to the right; you also get some skid to the left; but unless you are using your rudder viciously, the skid isn't even so very bad. For the airplane, having a built-in will of its own, will attempt to stop the skid by banking to the right. The nose drops slightly, and the air speed builds up slightly, but it seems hardly enough to bother.

So far, so good. If you relaxed your rudder at this point, you would get from then on a fairly nice shallow turn; you would simply have made a rough, skidding entry into that turn. But for the present experiment, keep on holding gentle right rudder, and observe.

You are now slightly banked; your right side is thus slightly earthward, your left side slightly skyward. When your rudder now swings your nose "around," it no longer swings it purely "around" along the horizon as it did when your wings were level; "around" now means also slightly earthward. Your nose goes down, your air speed builds up, and you begin to lose altitude. At the same time, the skidding continues, and the ship continues to try to stop the skid by banking up more steeply.

The more steeply it banks the more pronounced becomes the effect just described. By the time the bank has reached 45 degrees, your rudder swings your nose earthward fully as much **as it swings it** "around." At about this point you will almost certainly grab hold of the stick, because your air speed starts building up in earnest. If you still kept on holding rudder, this is what would happen: The ship would keep on banking up; the more it banked, the more "earthward" the action of the rudder would become. Pretty soon you would be in a steep, extremely fast corkscrew dive; and, if you still kept on (and had any altitude left), you would go past vertical into a slightly upside-down position, corkscrewing down.

This is bad enough and proves that what the rudder produces is not a turn, but something else. It is, however, only an experiment;

In a right bank, "right" is also "earthward," "left" is also "skyward." "Up," in the pilot's sense, means in the ground observer's sense also "somewhat to the right." That's why an airplane's controls do not work as one would expect at first glance, or even at second glance!

what actually happens in practical flying is less spectacular but much more dangerous. The pilot does not let a corkscrew dive develop. When he first notices how his nose swings earthward, the pilot starts putting back pressure on the stick to keep the nose up. He does this because he doesn't realize that he himself is ruddering the nose earthward; if he did, he would of course relax his rudder. Also, he suffers from the illusion that the "elevator" is the "up" control, and that "up" in this sense is where the sky is. In short, he is all confused— though happy.

All he actually accomplishes by putting back pressure on the stick is to force his airplane to higher Angle of Attack, slowing it up and bringing it closer to a stall. At the same time, the skidding kills some of his flying speed. Also at the same, time he now holds a lot of aileron to the high side (to the left if in a right turn), because he notices that the airplane wants to steepen its bank. A little of that overbanking tendency would be present even if the turn were flown correctly; but most of it is due simply to the pilot's holding right rudder, for the rudder causes a continuous skid to the left, and the ship responds to this skid by trying to bank to the right.

The pilot's aileron action, however, now causes an adverse yaw effect which swings the nose even more to the pilot's "right," which, because of the bank, also means earthward. This again, to the pilot's mind, increases the need to hold back pressure on the stick to keep the nose up, away from the ground.

It is in this fashion that the turn-by-rudder idea finally leads the pilot to barge around a right turn holding right rudder, at the same time holding the stick in the left rear corner of the cockpit: his controls all set for a spin!

Just exactly how does this disposition of stick and rudder invite a spin? Consider first of all what a spin is. A spin is nothing but a fancy stall; one side of the airplane is stalled, the other is not, and therefore the airplane sinks down twisting. Remember, next, that the direct and immediate cause of any stall is not "lack of speed," not "having the nose too high," but flying the airplane at too high an Angle of Attack. And the Angle of Attack at which an airplane flies is determined by the position, fore and aft, of the stick—the farther back the pilot holds the stick, the higher is the Angle of Attack. Thus, as the pilot gets the stick farther and farther back in his effort not to let the nose swing earthward, he is forcing the airplane to fly at higher and higher Angle of Attack and bring it eventually quite close to stalling Angle of Attack.

And now remember the tricky nature of the ailerons. The ailerons are devices to increase or decrease the Angle of Attack on the wing tips. As the pilot holds the stick well over to the left, in an effort to keep the ship from overbanking to the right, what he really does is to set the right wing tip to an extra-high Angle of Attack. But, seeing that the stick is well back, and the whole airplane is at high Angle of Attack anyway, this extra Angle of Attack on the right wing tip may well stall it; and if it does, the airplane spins off to the right.

Sometimes the story is a little more involved. It may be that the aileron does not immediately produce a stall but merely produces a very strong adverse yaw effect which slues the nose around even more sharply to the right, the "right" being also "earthward"; thus tricking the pilot into hauling the stick even farther back and stalling the airplane that way. If this happens, the airplane will possibly spin "over the top," that is, rear up momentarily, execute a "vertical reverse" and spin off to the left. But in any case, if a turn is made sharply enough under the turn-by-rudder formula, and continued long enough, only an extremely patient airplane could fail to spin.

Try it out sometime and convince yourself!

KICK IT INTO THE TURN?

The student pilot is no fool. As he makes his first turns under the coaching of an instructor, he discovers almost right away that his kid brother's turn-by-rudder theory simply does not work. In fact, this is one of the first big discoveries he makes about flying.

He finds that, once he has used ailerons and rudder in the beginning of the turn, both controls apparently become somewhat unnecessary. Once banked, the airplane seems to stay in the bank without having to be held in the bank by aileron! In fact, it wants to steepen its bank of its own accord and must be kept from doing so by a distinct and constant *opposite* aileron pressure, that is, leftward pressure in a right bank. Once turning, the airplane seems to keep on turning without much further need for rudder! In fact, he may sometimes catch himself flying a *left* turn (especially a climbing turn) with slight *right* rudder. It seems as if both stick and rudder could be centered and the ship would nevertheless keep on turning. He also discovers that during the turn he must hold back pressure on the stick to keep the nose from going down. This fact also was not in his kid brother's theory, and he can do nothing with it, mentally.

And finally he discovers that, at the end of a turn, when he wants to fly straight again, he can stop the turn only by using strong opposite stick and rudder—right rudder and right aileron to straighten out from a left turn.

All this the student discovers—or rather, he begins to discover; the trouble is that, when the facts don't fit our ideas, we usually try to disregard the facts—and especially so in the air, where it is often hard to study out just what the facts are, and where the facts are often rather odd.

And the way the rudder behaves in a turn is odd—if you think of this rudder as the turning control. It is as if your car, after turning a corner, kept on turning even though your steering wheel were straightened out again and as if it refused to straighten out from a left turn until your wheel were turned 'way over to the right. It simply cannot be quite true, the student feels.

Trying to hang on to the idea that the rudder is what causes the turn, here is how he usually figures things out: The ship is sluggish, he reasons. It takes a bit of an extra shove to set it turning. This extra shove is given to it by a rather pronounced kick on the rudder in the beginning. Once turning, however, the airplane tends to keep on turning because of that same inertia; and it would keep on turning more or less forever, unless you gave it finally a shove the other way around to stop it, this shove being given by the opposite kick on the rudder at the end of the turn.

That's the "kick-it-into-the-turn" theory of flying. It is a subdivision of the turn-by-rudder idea, a refinement of that vicious fallacy, rather than a fresh and sound idea; in this theory, the turning is still caused by rudder; only the manner and timing is different. The theory is popular not only among the student pilots but among pilots who ought to know better. Some instructor will no doubt protest that he doesn't advocate anything as crude as kicking. "You *never* kick," he says piously. "You exert a pressure." But that is only a question of vocabulary; the fact still remains that in his own mind he uses his rudder "to start the turn."

Well, you might say, maybe the reason why most pilots believe in it is that it works. Isn't it drawn from the observed facts? Isn't it the way the good, smooth pilot actually uses stick and rudder? True, one can fly for years by that formula and never get into trouble. But that's what makes it vicious. If a spin resulted from this theory every time, it would have been abandoned long ago. It survives because it works apparently well enough. But it is wrong, and it *will* cause trouble, especially on turns made in a hurry, suddenly, sharply, under mental stress—in an emergency. Remember, too many pilots die, and almost always by spinning out of turns. This is because so many pilots, when they make a turn, don't really know what they are doing.

How DOES AN AIRPLANE TURN?

Now that you have understood that the airplane is not turned by rudder, your mind may be free to accept the story of how an airplane is turned. This story is much more complicated and requires an energetic mental effort to understand it. Most people will rather die than think. This book is written for the exceptions, and so we start.

First of all, the formula: *An airplane is turned by laying it over on its side and lifting it around through back pressure on the stick.*

You have done it often, in steep turns. Most of us have gone through a stage where that was all the steep-turn technique we had—the well-known slam-bang system of doing "verticals" by which you first rolled over into steep bank, regardless of any slipping that developed while you did so, and then yanked back on the stick and pulled yourself down into your seat until you got woozy. This is a rough manner of making a turn and not approved, because it shows lack of "coordination." Still it is sound; it is based on the correct idea.

It is a pity, therefore, that we can't reverse the usual procedure and teach a student steep turns before shallow turns. Probably the heaviness and the crazy attitude would disturb him too much, and probably he would fly so erratically that the instructor would have to interfere too much. But it would give him the correct idea of an airplane's turn as no amount of explanation can.

To get the idea for yourself, go up sometime for a quarter hour and work your way gradually from steep turns toward shallow turns. What is so obviously true when you lie almost on your side and pull the stick back is still true at a 45-degree angle of bank. What is true in a 45-degree bank must also be true in a 30-degree bank, and so on down into the shallow turns. Feeling things out for yourself in that order, you never come up against a point where your mind demands that there must be some other cause (such as the rudder) to account for your airplane's turning; it is all bank, plus back pressure on the stick. True, as you make your turns shallower, the idea of laying yourself on your side and hoisting yourself around by pulling back on the stick become paler; but then, so do the turns thus produced; the turns become wider, less turning. If you lay yourself over only a little and put only slight back pressure on the stick, you get only a tame, gradual turn; and, vice versa, if you want a gentle turn, it is enough to use just a little bank and just a little back pressure on the stick. In no case, however, do you need rudder.

Why DOES AN AIRPLANE TURN?

Once you have got the feel of the turn, you may want an explanation: Why does the airplane come around?

The formula: *An airplane turns because its wings shove it over sideways, and its tail makes it weathercock.*

What makes the airplane come around is the force of its wings, more precisely, the force created on the airplane's wings by the impact of the air; the same force that also holds the airplane up. This force is commonly called "lift," but the word "lift" suggests to many pilots (wrongly) a force that always acts straight up. Actually, what engineers call lift does not necessarily act straight up. It acts at right angles to the Relative Wind and hence may act up or down or sideways or slightly forward—depending on the direction from which the relative wind strikes the airplane, and on the airplane's attitude. But we want to avoid here the rather barren sort of discussion that often passes as "theory of flight" and doesn't really explain things, but merely gives them names, preferably Latin ones. For this discussion let's call this force simply the wing force, and let's make the simplifying assumption that it always acts perpendicularly to the wing.* Thus, if the wings are level, the wing force pushes straight upward. If the wings were upside down (as on top of a loop) the wing force would act straight down toward the earth. If the wings were exactly up and down in a truly vertical bank, the wing force would act to push the airplane over toward the horizon.

In straight flight, then, when the wings are level, the wing force simply holds the airplane up against the pull of gravity. But if you bank, you thereby tilt the direction of the wing force. With the airplane tilted over to the right in a 45-degree bank, the wing force acts no longer straight up but is also tilted 45 degrees over to the right. Hence it no longer merely holds the airplane up, but it also shoves the airplane over to the right.

This alone would not produce a turn. If no other force went into action the airplane would then simply slide sideways through the air over to the right. What makes it execute a proper turn instead of such

---

* An engineer will object that the air force acting on a wing does not necessarily act perpendicular to the wing. But he will also agree that for the present discussion this assumption introduces no error.

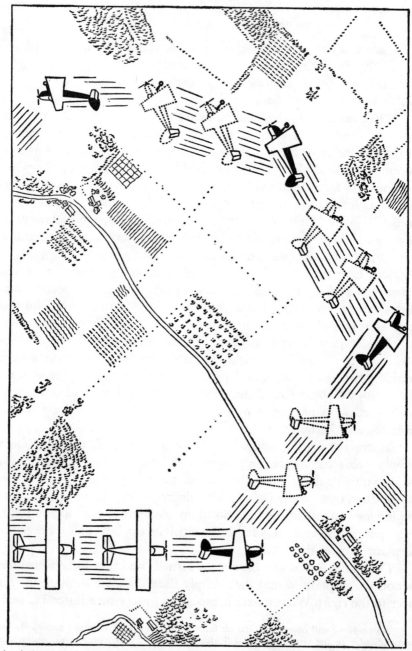

An airplane turns because it is banked. The view here is vertically down. The airplane is not diving or looping, but is flying a level turn at about 45 degrees of bank. Explanation on opposite page.

sidling motion is the impact of air upon the ship's tail. The tail, acting much like the tail feathers of an arrow, won't let it slide through the air sideways, but makes it swing around, much like a weather vane, nosing the airplane around into the Relative Wind produced by its own motion. When the wing force shoves the airplane sideways through the air over to the right, the tail is blown to the left, and the nose thus swings to the right; the airplane "turns into the bank."

This is, of course, an "exploded" account; it takes a number of events that happen simultaneously and pulls them apart into a step-by-step time sequence. Actually, the wing force starts pulling the airplane sideways from the first moment on when the airplane is even the least bit banked; and actually, the sideways sliding of the airplane can never develop, because the tail will never allow it to develop: the tail begins to weathercock the airplane around as soon as there is even the merest tendency of the airplane to move sideways through the air. It is not necessary that there be actual sideslip before the airplane will turn, after it has been banked. The turning follows from the bank exactly because the airplane is shaped so as to *avoid* sideslip!

But now to continue with our account of the turn. Now that the airplane has turned, its wings are still banked. Its wing force still pulls it sideways, sideways as counted from the new direction of flight; again it would slide sideways through the air, and again the sideways Relative Wind that results from the sideways motion blows the tail around and causes the airplane to turn. In this fashion, a smooth steady turn results from the bank.

Explanation: this picture shows in separate rough phases a process which is really continuous and smooth. The style of motion indicated by the dotted airplanes never really happens, it is merely continually threatening to happen, and the airplane's turn results from its continuous attempt to avoid this style of motion. *The airplane behaves as the solid airplanes are behaving in order to avoid moving as the dotted airplanes are moving.*

At station 3 the airplane is banked. The lift of its wings then pushes sideways as well as upward, and makes it move in the direction of station 6. If no other forces were acting on it, the airplane would move from station 3 to station 6 in the sidling fashion shown at stations 4 and 5. Such sidling motion is prevented, however, by the weathercocking action of its tail, which always tends to turn the airplane's nose into the Relative Wind. This nosing-around is shown at station 6. Once nosed around (station 6), the airplane is still banked, and its wings still push sideways, and thus the whole process keeps repeating itself as long as the airplane is banked.

Thus you can see that an airplane turns because it is banked and for no other reason. It needs no rudder. And you can see even now, before going more deeply into the matter, a most important rule for pilots: If you want a sharper turn, there is only one way to get it— not rudder, or "more rudder," as so many student pilots believe, but *bank:* more bank.

The steeper the bank is, the more sideward does the wing force pull; hence, the more lively is the turn. This should be precisely understood by every student. The student tends to think that the turn is caused somehow (probably by rudder) and that he banks the airplane merely to make himself comfortable—much as a driver on a curve produces the turn by twisting his steering wheel and at the same time wishes the road were banked, because that would make it safer and more pleasant. In short: to the student it seems as if the bank were a consequence of the turn, but that is wrong; the turn is a result of the bank! *The more of a turn you want, the steeper must be the bank.*

This should be so firmly etched in a pilot's mind that he will remember it in an emergency—expecially when maneuvering for a forced landing after engine failure. In such an emergency there is always a tendency in the first place to hold the stick too far back, in a confused attempt to conserve altitude and "stretch" one's glide. This makes the airplane vulnerable to any further misuse of stick and rudder. If the pilot then tries to turn sharply, perhaps in order to make a certain field, there is always a tendency to try to hurry the nose around by rudder, especially because there is also a natural reluctance to bank steeply while near the ground. Such an attempt to keep the bank flat and hurry the turn by rudder is likely to be the last thing the pilot ever does.

WHY "BACK PRESSURE" IN A TURN?

But what about the back pressure on the stick? The back pressure on the stick is not necessary in order to get a turn; it is necessary only in order to get a good turn and to avoid a loss of altitude during the turn. The smoothness of the turn depends very much on correct application of that back pressure. The student usually worries a lot about the exact coordination of rudder and aileron; what he ought to worry about is the coordination of back pressure and bank. All this we shall now try to make clear.

First of all, the why of it. What happens if the pilot simply banks the airplane without putting back pressure on the stick? The wing force, which in straight wings-level flight acted straight up and was just big enough to hold the airplane up, is now tilted and has a double job; part of it goes to hold the airplane up; part of it goes to shove the airplane sideways. But the wing force is no bigger now than it was in straight wings-level flight; therefore, it cannot do either of the two jobs quite to satisfaction. It now fails to some extent to hold the airplane up; the airplane sinks, and in sinking noses down. At the same time, the wing force also fails, to some extent, to push the airplane sideways as energetically as it should. The resulting turn is therefore sluggish, and the airplane slips off to the side toward which it is banked.

This is not dangerous and would never lead to loss of control. In fact, the airplane will correct for it by its own stability—its own built-in will-to-fly. Disregard the slipping, the dropping of the nose, the sluggishness, the loss of altitude for a few moments, and the airplane will pick up speed. As the speed builds up, the wing force builds up—the force is made, after all, by the impact of air upon wing, and that impact increases "as the square of the air speed." If the speed increases even only a little, the air hits the wing very much harder. As the wing force grows big enough for its dual job, the airplane finally stabilizes in a beautifully flown downward spiral, flying at high air speed and losing altitude, but making a perfect, nonslipping, not at all sluggish turn.*

Ordinarily, however, we don't want to fly downward spirals, but level turns. We sometimes can't afford loss of altitude in a turn, being perhaps too close to the ground. Also, we prefer not to slip or skid. And that's why back pressure must be held against the stick in every level turn; the wing force must be increased as the airplane is banked. By forcing the stick farther back, we force the airplane to fly at higher Angle of Attack; for the stick is the airplane's Angle-of-Attack control. At the higher Angle of Attack, the wing then produces a bigger force; and the bigger wing force is then adequate for the double job of holding the airplane up and shoving it over to the side. A level, non-slipping turn results.

The steeper the bank, the bigger is the additional job that the wing force must do—additional to its original job of holding the air-

* Many pilots will tell you this isn't true. You try it.

plane up. It follows that the steeper the bank, the more back pressure is needed on the stick. In order to get a smooth and level turn, therefore, the amount of back pressure the pilot applies on his stick must always be exactly "right" for the angle of bank that his airplane happens to have.

A PROBLEM IN COORDINATION

Thus the how of back pressure is also worth studying. This is where the student tends to neglect proper coordination of controls. In the beginning of the turn, while still steepening the airplane's bank, he tends to wait with his back pressure until he has the airplane banked as he wants it; and only then does he apply back pressure. This is the above-mentioned slam-bang system of making turns, and it is rough. For its result is lack of back pressure during the beginning phase of the turn. This makes that beginning phase sluggish and slipping. If a pilot still has turn-by-rudder ideas in the back of his head, he then tends to remedy the trouble by using bottom rudder. This remedy, however, is worse than the trouble!

To get a perfect entry into the turn, the back pressure must be applied and gradually increased even as the airplane rolls into the bank; *at any one moment* the back pressure must be exactly right for the steepness of bank that the airplane has reached *at that moment*.

This is difficult to do. What makes it so difficult is that the pilot is accustomed to coordinating control pressures with control pressures; he is accustomed to accompanying such and such a hand pressure against the stick with such and such a foot pressure against the rudder. But the firmness of back pressure on the stick during a turn must be governed, not by any other control pressure the pilot is exerting at the same time, and particularly not by the sideways pressure he is exerting on the stick while rolling the airplane into the bank. It must be governed by *a visual impression:* how steeply the airplane is banked, as judged by the way the horizon seems tilted in front of the ship's nose. Roughly, he can also judge by the way the nose behaves in the skyward, earthward sense: he must apply enough back pressure to keep the nose from going down. (In fact, the nose appears noticeably higher, relative to the horizon, during a correctly flown steep turn than it appears in straight and level flight; this is because of the much higher Angle of Attack during the turn.)

ERRORS AND CORRECTIONS

It may now be worth while to discuss certain errors that the pilot is likely to make in a turn and certain corrections he can make.

If he applies too much back pressure, what happens? The wing force is then excessive, and it will carry the airplane around briskly and at the same time lift it up. The ship will carry the nose rather high and will gain altitude. But there will be no slip or skid. The ship will simply execute a perfectly flown climbing turn. This can also be stated the other way 'round: If your altimeter creeps up during a turn the reason is that you are holding too much back pressure, and the remedy is to relax your back pressure slightly.

If the pilot applies too little back pressure to the stick, what happens? The airplane slips slightly, at least for a while, until the speed has built up. It carries its nose low and keeps losing altitude. A pilot who has turn-by-rudder ideas is then tempted to use bottom rudder in an attempt to get more turn and stop the slip. But the remedy is simply to increase the back pressure slightly. This also can be expressed the other way 'round: If during any part of the turn the airplane slips, the reason is almost certainly la'k of wing force, that is, lack of sufficient back pressure. The rudder should be left alone until an increase in back pressure has been tried. If an airplane loses altitude during a turn, the reason is again lack of sufficient wing force, and the remedy is to increase the back pressure. This is so in a correctly flown turn. If the ship is losing altitude because it is being stalled, that is another matter.

The secret of keeping a turn strictly level lies therefore entirely in the proportion of back pressure to bank. For a given bank, too much back pressure gains altitude, and too little back pressure loses altitude and temporarily also causes slipping. Instead of adjusting the back pressure to the bank, the pilot may also do the opposite, change the bank so as to fit it to the amount of back pressure he chooses to hold. For example, he might be in a very steep turn, holding a great deal of back pressure; yet he may notice that his altimeter is creeping down. To put on even more back pressure seems somehow wrong to him—and rightly so—it would be likely to stall the airplane. Thus he simply keeps the back pressure constant and takes off a little bank— almost unnoticeably little will do the trick. This restores the proper proportion between bank and back pressure and stops the loss of altitude. Or, if the pilot takes off still another bit of bank, while holding the original back pressure, he gets a slightly climbing turn.

All this is true of shallow and steep turns alike; except that, in the shallow turns, the pilot will usually prefer to adjust back pressure to bank, while in steep turns he will often prefer to adjust bank to back pressure.

Notice that the rudder has no function in all this. There is a theory still ghosting around the older airports that in a steep turn "the controls reverse their functions, and the rudder becomes the elevator," that is, that in a steep turn you regulate climb and descent and the position of your nose above and below the horizon by giving top or bottom rudder. There is no difference between the steep turn and the shallow turn, as far as control functions are concerned; and the less you think about this theory, the better off you'll be.

THE LAWS OF THE TURN

The airplane's ability to turn quickly and sharply, often called its *maneuverability*, concerns not only the fighter pilot. It concerns the pilot of every ship in all maneuvers. Every small turn the pilot makes, every smallest correction of his flight path, is subject to the same laws that also govern his sharp, tight, steep turns. (These laws are closely tied up with the whole matter of back pressure and wing force, g load, and persistence in straight motion.) We shall not show the mathematical connection here but shall simply state what these laws are and how they affect the pilot.

Turning ability depends on three factors. By far the most important is speed.

The speedier ship needs very much more room to turn around in. If you fly twice as fast, you need, at any given angle of bank, four times the room.

For example, at a medium bank, a 100 m.p.h. airplane may need about 1,000 feet to turn around in (turning radius 500 feet). A 200 m.p.h. airplane, turning at the same medium bank, needs 4,000 feet to turn around in! When a pilot transfers to a faster ship, this is at first likely to spoil all his planning. He overshoots all his turns, he is carried too far out, or else he finds that he keeps having to bank up much more steeply than he had intended. But that isn't all.

The speedier ship also needs more time to turn around in! If you fly twice as fast you need (at any given bank) twice as much time to accomplish a given change of direction. For example, the 100 m.p.h. ship can reverse its direction of flight by holding a medium bank for half a minute; a 200 m.p.h. ship will need a whole minute to reverse its direction of flight at that same angle of bank; if it wants to turn around in half a minute, it must use a much steeper bank and much more back pressure and must fly much closer to the stall.

This effect often startles observers on the ground; when a very fast airplane makes a turn, it is often in a steep bank and yet seems to be flying almost straight. You would swear that the pilot is flying on his side in a slipping turn. A glider's turn, on the other hand, often looks "skidded"; the glider is banked only moderately, but yet it swishes right around. Actually both ships are presumably flying proper turns; proper turns, flown at such different speeds, will simply have such different appearances, according to perfectly clear laws of the turn. So precisely do those laws work out that you can calculate the speed of an airplane if you can observe only two things, the angle of its bank, and the time it takes to accomplish, say, a 90-degree change of flight direction.

A pilot notices this effect very much when he first transfers to a much faster ship. The faster ship seems "stiff." He puts it into a reasonable bank and applies back pressure, and it feels all right, but the nose doesn't seem to come around as quickly as he thinks it should. He tries on a cross-country flight to make a small course correction, say 5 degrees; and if he wants that correction at all promptly, he finds that it takes quite appreciable bank to get it.

It is often said that, in a very fast airplane, even a slight turn will lead to the pilot's blacking out—owing to excessive g load. This is misleading. For any given angle of bank, the g load on the pilot is the same, regardless of the speed, power, weight, wing loading, or size of the ship; in a light plane or a fighter, a 60-degree bank will make you twice as heavy as you are. But understood in another sense, the statement is true: It a turn is to be accomplished in a prescribed time or a prescribed space, the fast airplane may black out its pilot. Suppose that both a light plane and a fighter were following a railroad track with a curve in it. The light plane could take that curve with only a medium bank and would not feel much g load. The fighter, owing to its high speed, might have to bank up to 80 degrees to take the same curve—and at about 80 degrees the g load is 5.75, and blacking out begins.

The second factor that affects turning ability is the airplane's ability to maintain a very steep bank without slowing up so much, owing to increased wing drag, that it will stall. This depends mostly on the power available, but also on wing shape. When flying at high Angle of Attack, the long narrow gliderlike wing has less drag than the short, stubby wing of the same area.

SLOW TO ROLL

The third factor that affects turning ability is the quickness with which the airplane can assume a bank, and hence a turn. At first glance this seems to depend mostly on the size of the rudder, and one sometimes hears arguments to that effect. Actually, the rudder does not produce the turn; and as long as it is big enough to take care of the adverse yaw effects of the ailerons, even with the ailerons hard over, it is big enough; and enlarging it or giving it more throw will not enhance the ship's maneuverability. The effectiveness of the ailerons is much more important, for it is the rolling into the bank that consumes precious quarter seconds. Even with overwhelmingly powerful ailerons, however, you simply cannot roll an airplane into a bank instantaneously, because the wings themselves resist any attempt to roll the airplane fast.

This is called the *lateral damping effect* of the wings. When the airplane rolls into a bank, one wing is forced to squash downward through the air; and, the faster it tries to squash, the more firmly the air resists it. At the same time the other wing tip must squash its way upward through the air; and again, the faster it squashes through the air, the more firmly the air resists; thus the rolling of any airplane is strongly "damped." This effect is of course the stronger the longer the wing; because, the farther out the wing tips are, the more actual up or down travel they must do for any given rolling motion of the ship; and hence, the bigger is the resisting force. In addition, the resisting force then acts on a longer lever arm; then, if you double the wing span, you quadruple this damping effect. That is one reason why in the old days pursuits were biplanes or even triplanes; it made them quicker to roll into a turn.

The effect should not be misunderstood. It does not mean that some airplanes can't bank steeply; it merely means that some airplanes can't achieve a steep bank very fast. Simplifying enormously,

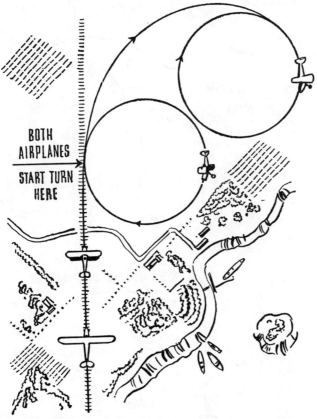

BOTH
AIRPLANES
START TURN
HERE

"Damping in roll:" Two airplanes of the same speed, flying at the same angle of bank, will always make turns of the same radius. But they may differ in another respect. A short-span airplane can whip into a bank and can thus establish a turn almost instantly. A long-span airplane is slow to roll into a bank, and hence will be carried out farther before the turn is established.

take a monoplane and a biplane, both of the same speed, flying abreast of each other down a railroad track; both will have the same turning time, and the same turning radius. But both are not equally quick to start a turn. Suppose that at a given signal both are required to "stop," that is, go into a steep continuous turn. A few moments after the signal the biplane is steeply banked and turning tightly while the monoplane has as yet achieved only a shallow bank and a wide turn and is still trying to steepen its bank and tighten its turn. The result is that the monoplane will be carried farther out and over to the side. (Because of their high wing loading, consequent short span and high

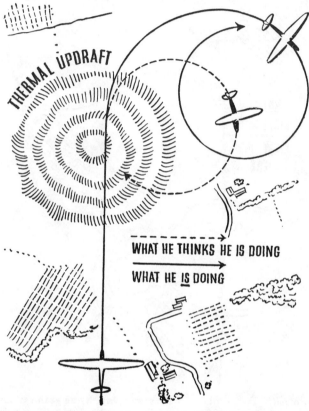

The soaring pilot fails to find the updraft again after flying through it once. Because of its long wing span, the glider is slow to roll into a turn, and what the pilot intended as a circle becomes a figure six.

speed, modern fighters have no excessive lateral damping effect even though they are monoplanes.)

This effect is especially noticeable in gliders because of their large wing span. You cannot roll a glider into a turn faster than a stately, gradual leaning over; even with the ailerons hard over, it simply won't roll any faster!

THE GLIDER MISSES THE BUS

This is troublesome to the soaring pilot: you glide along straight, hoping to find an updraft. When you finally hit one, it is usually so small that you fly right through it. Your problem then is to get back into that same spot of air. In first-sight theory you need now merely to assume a bank—any bank—and hold it constant, and the resulting circle cannot fail to bring you back to the updraft spot. But in some gliders it takes so long to establish a bank that the intended circle becomes a figure six, and brings you out beside the updraft spot; you have missed the bus!

THE UPWARD TURN

Everything that is true of the turn, of the curving of the flight path sideways, is true also of the pull-out from the dive, the flare-out from the glide, the pull-up into a loop—in short, of any curving of the flight path upward. An upturn of that kind is in most respects simply a turn of zero bank. It is subject to all the laws of the turn.

This sounds like a theoretical and useless idea, but it isn't. It is practical and interesting.

It warns you, first of all, that you can stall with your nose down in too sharp and upward turn just as you can stall, with your nose on or below the horizon, in too tight a sideward turn. Even in a straight-down dive, even at terrific speed, just get the stick back far enough, and you will stall and spin—unless the excessive wing force or excessive g load (same thing, really) breaks the wings off first, which in civilian ships is much more likely.

You can say such a stall is caused by the excessive g load, or you may say it is caused by too backward a position of the stick or by too high an Angle of Attack. The three things necessarily go together; the stick position causes the Angle of Attack which produces the greatly increased wing force which causes the deflection of the flight path which causes the g load which makes the wing force necessary! It is all the same thing—simply a turn.

NOSE-DOWN STALLS

This type of nose-down stall occurs sometimes at the end of a loop; the student has felt light in his seat on the top of the loop and has hauled the stick all the way back trying to tighten the loop, to keep himself pressed into his seat, and also to hurry the maneuver to a finish. On top of the loop, while the airplane is upside down, the ordinary laws don't quite apply. But at the moment that the nose is vertically down again, the normal laws apply again and—the stick being all the way back—the airplane stalls right there, with its nose pointing straight down, and spins.

Another example of such a nose-down stall, caused by an upward turn, happens sometimes when an upward turn is combined with an ordinary sideward turn. Suppose that, during a turn which was intended to be level, you have allowed your nose to drop and have picked up a lot of speed. You now try to get the nose up again where it belongs simply by applying more back pressure. This is perfectly sound flying; it will do the trick and will do it without slipping or skidding. It is in fact the only way in which you can do both—keep on with your turn and stop the loss of altitude. (The only other way to get the nose up would be to shallow the bank, widen the turn, and keep the back pressure constant.) But it means that you are doing two turns, one upward, one sideward, both at the same time. You exert back pressure enough for two turns; you get g load from two turns at once; you get Angle of Attack enough for two turns at once; you may easily stall (with the nose well down, mark well, and in a quite moderate bank as well, and at perhaps at very high speed). Or here again, the excessive wing force, alias excessive g load, alias excessive back pressure, might take off your wings.

DIVING ON GIRLS' HOUSES

Next, the upward-turn idea is of practical interest for the poor suckers who dive on girls' houses. *For the same laws that govern the sideward turn also govern the upward turn in regard to the time and the room needed to turn.*

In order to regain level flight out of a vertical dive at 100 m.p.h., you need such and such an amount of room, depending on the g load you are willing to put on yourself and on your airplane—say you need 400 feet. But out of a dive at 200 m.p.h. (in the same ship, or in any other ship) you need 1,600 feet for recovery to level flight, provided that you want only the same g load as before! This means that every second during which you are in steep dive, picking up speed, not only kills a lot of altitude but adds enormously to the amount of *additional* altitude you are going to use up when you want to stop the dive!

It is simply math, but its cumulative effects are perfectly astounding—and frequently fatal. First, in a big way, every once in a while

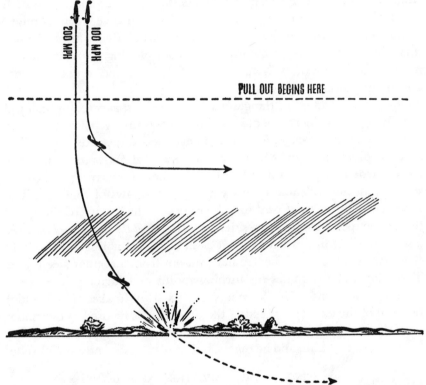

At twice the speed, you need four times the room to accomplish a turn. That goes for turns in the up-and-down sense no less than for right-and-left turns. Here both pilots are diving straight down. Both start pulling out at the same height. Both subject themselves and their airplanes to the same *g* load—perhaps 5 *g*'s. At 100 m.p.h., recovery is possible with plenty of room to spare. At 200 m.p.h., recovery is impossible—if the pilot tried to pull out harder, the *g* load would pull off his wings or would make him stall and spin.

some pilot grossly underestimates the amount of room he needs for the pull-out and dives steeply into the ground. Or, realizing his predicament, he pulls back so hard that the wings collapse; or, he pulls himself into one of those nose-down stalls, and spins; or, perhaps, he blacks himself out while quite near the ground, and hits something.

Beware, therefore, of your first few flights in a new airplane which is faster than what you have been flying and probably also cleaner, so that in a dive it picks up more speed faster. Such an airplane is likely to have retracting landing gear and possibly also an arrangement by which the throttle can't be closed unless the wheels are down. In such an airplane it is hard to get out of a high-speed dive, such as might result from stunting or from loss of control while flying blind. Hence it is wise to do any experimenting with your landing gear extended—in order both to reduce the cleanness of the ship and to enable yourself to cut power.

But this effect—the dangerous increase in space requirement if the diving speed increases only a little—is positively poisonous when it works on a small scale; for it is then even less suspected by the pilot. Shallow diving at a girl's house, a pilot underestimates the amount of room needed by 20 feet and hits the chimney; the error of judgment seems quite inexplicable unless you know the laws of the turn. The room needed for an *upward* turn increases as the square of the speed just as the room needed for a sideward turn. *And the time needed increases also!* The *slight* upward turn from a shallow dive is just as subject to that law as is the more spectacular pull-out from a regular steep dive. The pilot underestimates his turning radius and turning time in the up-and-down sense. He "overshoots" his turn (makes it too wide) because of the excessive speed of the dive, just as he used to overshoot his right-and-left turns the first time he flew a faster ship. The chimney looms in front of him, and he reacts, but he reacts too gently and too late.

WING LOADING AND THE LANDING

But the most interesting application of the turning laws to the upward turn is this: They explain the difference between piloting a heavily wing-loaded and piloting a lightly wing-loaded airplane, especially on landing. It may be of interest to anyone who plans to transfer from slower to faster more heavily wing-loaded equipment.

Compare a light airplane, wing loading 6 pounds per square foot, with a cargo ship, wing loading 24 pounds per square foot, and assume that both ships are equally well behaved. (This assumption is not unreasonable. Airplanes of different wing loadings usually do differ also in control feel, in stability characteristics, in stalling characteristics; but that is not because of their different wing loadings but because of the different purposes for which they are designed. There are heavily wing-loaded airplanes whose stall behavior or whose control feel is much like that of a light plane. The discussion here will disregard other differences between fast and slow ships and concentrate on the effect of wing loading alone.)

Other things being equal, then, the cargo ship will make its approach and landing at twice the light ship's speed. This much based on the physical laws of lift. But this means that the heavy fast ship makes all turns, for a given g load, four times as wide! And that goes not only for turns right and left, but also for upward turns, particularly, the flare-out for the landing. In the light ship, you are accustomed to flare out at such and such a height and while still such-and-such a distance from your intended landing spot; in the heavy ship, using the same landing technique you would have to *start flaring out four times as high, and four times as far away!* If you didn't, you would have to pull up more sharply, possibly so sharply that the g load would squash you right through on the ground.

But that isn't all. Not only is the spacing different, but also the timing: since he is flying twice as fast, the pilot of the faster ship needs twice the time and *must begin his flare-out twice as many seconds before the actual landing as the slow ship pilot must.*

TIMING AND SPACING

Note that both the time factor and the space factor tend to trick the slow-ship pilot into flying the heavily wing-loaded airplane into the ground or else having to pull it up very sharply at the last moment. Experience shows that this is indeed the way the green pilot behaves on a heavy ship. And note that both the time factor and the space factor force the pilot to start his landing maneuver while he is still well away from the ground and hence not able to judge with as much precision as he can closer to the ground. Thus the heavily wing-loaded ship poses two problems to the pilot. First, he must get accustomed to a new rhythm in timing and spacing. Second, even then he needs more accurate, more skillful sensing of height, flight path, speed, and all other factors that make or break a landing. (This discussion assumes that both pilots are making the same *type* of approach and landing, for example, both are making three-point landings, or both are making wheel landings.)

But even that isn't all! Heavily wing-loaded ships usually glide very much more steeply than do the light ones. This is not, as many students think, a direct result of the high wing loading. It is a result rather of the enormous power plants that necessarily go with high wing loadings; when idling, those motors, nacelles, and propellers act as enormously powerful drags. Because of this steeper glide, the heavily wing-loaded ship has more upward turning to do in the flare-out. This makes it necessary to flare out even earlier, even higher—in

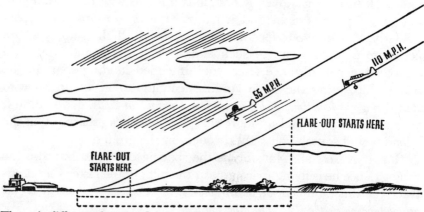

The main difference between flying a light plane and flying a heavily wing-loaded ship is in the landing flare-out. The heavier ship's normal glide is faster; and because the turning laws apply to upward turns just as to right-and-left turns, the pilot of the heavier ship must start his flare-out much earlier in *time*, and very much higher and farther back in *space*.

order to avoid having to flare-out even more sharply, with more *g* load. This is one reason why heavily wing-loaded ships usually do not make the same type of landing as lightly wing-loaded ones but make their landing approach with power on, along a shallow path of descent. This technique makes it easier for the pilot in two ways; not only is the approach more accurate, but also there is less upturning to do at the end.

THE NASTY SMALL-SCALE EFFECTS AGAIN

When a pilot dives on his girl's house, the small-scale working of the turning laws trying to make him hit the chimney is more treacherous, perhaps, than their big-scale working, trying to dive him into the ground at a steep angle. The same thing is true in landing the heavily wing-loaded ship; the small-scale working of the turning laws is most important. Not only the flare-out itself, but *every small correction is also subject to the turning laws;* the ship is "stiff" in the up-and-down sense just as it is "stiff" in the right-and-left sense. In landing a light trainer, most pilots make many small quick fore-and-aft motions with their stick, making small corrections, up or down, of the flight path; and in such a lightly wing-loaded ship these corrections take effect practically instantaneously as to time and within inches as to height. In the more heavily wing-loaded airplane, landing twice as fast, the *same corrections need four times the room and twice the time to take effect!* The airplane reacts within half seconds or seconds as to time and within feet as to height. Thus the heavy-ship pilot cannot make a quick correction so easily; he must therefore use more accurate judgment to begin with.

SUPERMAN?

To sum this up, then, there is no "real" difference between the heavily wing-loaded and the lightly wing-loaded airplane. Both are the same sort of machine, subject to the same laws. But so are a flying baseball and a flying bullet the same sort of device, subject to the same laws. For the human being concerned, the same laws produce entirely different results. The lightly wing-loaded airplane can be handled in the ordinary working style of ordinary men, including the writer's, which is a process of fumbling, of trial and error. The pilot simply watches, finds out what he is doing, and then makes a correction. The heavily wing-loaded airplane demands more precision; the pilot cannot watch for imperfection to become apparent, but he must be able to anticipate what the ship is going to do, he must more nearly really *know* what he is doing.

THE USE OF THE RUDDER IN TURNS

And now, once more, the rudder: What does the rudder do in a turn? This has already been discussed in the chapter concerning the rudder. It was shown there that, as you roll the airplane into a bank, the ailerons cause an adverse yaw effect which must be counteracted by rudder. The same thing is true on rolling out of a turn; an adverse yaw effect appears which calls for the use of rudder. On rolling out of the turn, more rudder is needed than on going in. This is because during the turn the pilot has forced the airplane to higher Angle of Attack (in order to get sufficient wing force to accomplish the turn); and at high Angle of Attack the adverse yaw effect of the ailerons is more powerful; hence the rudder action must be more powerful.

Thus the rudder does go into action during the turn. But it does not *cause* the turn. Someone may object that, in instrument flying, we regulate rate of turn by rudder. And instrument flying is scientific flying. Anything we do "on instruments" is probably correct.

This is half true, half false. It is true that instrument flying is more scientific and that the use of the controls while "on instruments" is apt to be more correct than while we are flying by outside vision. It is also true that we react to the position of the turn needle by rudder. But it is not really true that we regulate the turn by rudder. We do so only in a make-believe way—we act "as if" the rudder regulated the turn. Actually, when we use the rudder to make the needle point as we want it to point, we don't turn the airplane. We merely *skid* the airplane, thus causing the ball to roll to the high side; and thus we force ourselves to use our ailerons to bank the airplane—and the bank then causes the turn! And to this writer it seems highly dubious that the present rule—control of the turn indicator by rudder—will survive much longer. It is not scientific, not correct, and not efficient. It is a blemish on our technique of instrument flying. And, as a matter of fact, it is frequently disregarded in practical instrument flying.

The rudder does nothing in a turn that it does not also do in straight and level flight. In *all* flying, the rudder's essential function is to keep the airplane from yawing, to counteract all disturbing influences, whatever their source, that would tend to slue the airplane around and make it slither sideways through the air.

The two biggest of these disturbances are the adverse yaw effect of the ailerons, and the so-called "torque." And it so happens that in a turn both these disturbances are comparatively plentiful. At the beginning and the end of the turn, while the airplane rolls into the bank and out again, there is much use of aileron, and during the turn itself (especially a steep one), the power plant labors heavily while the air speed is somewhat reduced, and there is much torque. Hence the rudder is comparatively busy when the airplane makes a turn. But still it neither produces the turn nor helps produce it; it neither stops the turn nor helps stop it. Bank and back pressure do that.

"But how am I to know," a student might ask, "whether to use rudder at any given moment, and on which side, and how much?" There are two answers to this.

First of all, you know it by reasoning—if you understand the purpose of the rudder. Whenever you put rightward pressure on the stick you know exactly, "This must now be causing some yaw to the left. Quick, Henry, the rudder!" And when you go around a steep turn with wide-open throttle and well-back stick you know, "This now must be causing a lot of torque. I'll have to hold against that with my right foot." It is amazing how comparatively well you can fly that way—"mechanically." Flight instructors are usually horrified at the idea, but it is a fact that control coordination is fundamentally a mechanical problem and can be achieved by "mechanical" flying. Remember that "two-control" safety airplanes fly beautifully that way, by a perfectly mechanical linking of stick and rudder, really of aileron and rudder, so that they can always only move "both together." Of course, if you want to fly mechanically, you must understand the mechanism you are working; you certainly cannot fly mechanically if you don't understand your airplane. In this instance, you must of course rid yourself of any last hidden subconscious trace of the idea that you need rudder if you want to turn or that you will get a turn if you use rudder; the important thing is to make the rudder entirely subservient to the aileron, to put on a touch of rudder whenever you put on a touch of aileron, however slight and to take off your rudder pressure whenever you take off your aileron pressure on the stick.

There is another answer though, one that finally throws you back on that time-honored sense organ of all pilots—the seat of your pants. Whenever you feel as if you were about to slide in your seat toward one side, it is a sign that something is yawing the airplane, and the rudder is needed. You need not know what that something is—aileron yaw, torque, or maybe some other effect—just put on rudder on the side toward which you are being thrown. For instance, in a turn to the left, you may have a feeling that you are sliding toward the left in your seat; put on slight left rudder and the feeling will disappear. A little later in the same turn you may have the feeling that you are being thrown against the right-hand side of the seat; put on some right rudder; never mind that it is a left turn; put on right rudder until the feeling disappears.

"Well," you might say, "that's what my kid brother told me in the first place." But it isn't; for this is true only provided that you are flying a basically correct turn; particularly, provided that you are carrying the correct amount of back pressure on the stick. If you don't, if you just bank to the right without putting on back pressure, you will get a slip to the right, and to correct that slip by rudder would lead to undesirable results—as we have seen.

Thus the rudder, in the turn and in any other condition of flight, can also be described as a balancing control; whenever the pilot feels off balance, he uses the rudder to restore his balance. Now the pilot does not feel off balance unless the airplane is skidding or slipping, that is, slithering sideways through the air. He will feel off balance every time the airplane does skid or slip. And, putting it still another way, every time the pilot feels himself pulled sideways in his seat, the airplane is being pulled sideways through the air—inevitably so, if you think about it. To use the rudder for balancing the feel of the seat of your pants means therefore exactly the same thing as using the rudder to counteract yawing effects!

The Wright brothers, who knew just about everything about the airplane, knew this, too. As early as 1910, in the course of their famous lawsuit against Curtiss, a foreign patent court declared that part of their original invention was that the rudder was not a turning control but a balancing control! There is no excuse today for the pilot who still thinks that the rudder is what turns the airplane.

HOLDING RUDDER DURING A TURN?

And now, for grown-ups only, a special little treatise on something which it isn't good for the young to know—how sometimes you do hold inside rudder in a turn after all. It has been emphasized that the rudder does not produce the turn; this is completely, and without exception, true. It has also been claimed that the rudder does nothing in a turn which it does not also do in straight and level flight. This is not always exactly true in all airplanes. In some airplanes the turn produces a disturbance which must be counteracted by holding inside rudder.

Imagine an airplane, otherwise normal, whose vertical tail fin is mounted on a tail boom a mile long. Let this airplane do a 45-degree banked turn to the right. Obviously, that tail fin would behave much as a long pole behaves that sticks 'way out behind the rear end of a truck: when the truck turns a corner to the right, the end of the pole does a wide vicious sweep to the left. Such a far-extended tail fin, then, would in a right turn sweep around in a wide arc to the left, pressing with its left side against the air; obviously, the air would resist its sideways passage and would slow up its sweep. The fin would

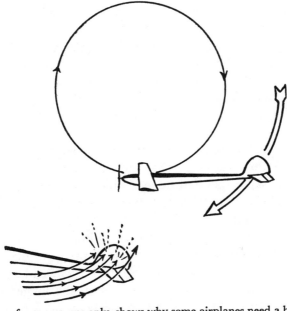

This caricature, for grown-ups only, shows why some airplanes need a bit of "inside" rudder after all during a turn. If an airplane's tail is long enough and its speed slow enough, so that its turning radius is short, its vertical tail fin meets a Relative Wind somewhat as sketched in the lower left corner. The tail then cannot swing around as quickly as would be required for a slipless turn, and must be helped around by a little "inside" rudder.

keep the airplane from turning as fast as it should, considering the bank, and the airplane would slip. And, if the pilot of that ship wanted a proper turn, he would have to help the tail fin swing around by using right rudder.

This condition, "damping," occurs also on airplanes of normal shape. In fact, it is more frequent than we usually realize. Consider the evidence: In all airplanes in a medium steep turn, we hold the stick to the high side against the overbanking tendency; and we could therefore expect, according to what has been explained so lengthily throughout this book, to have to hold a lot of top rudder as well, in order to counteract the adverse yaw effects of those ailerons. But, if you watch yourself, you find that in many ships you do not in fact hold such top rudder in a turn. You find that there is no need for it. Well—why not? Who, or what, is taking care of the adverse yaw effect?

The answer is that this sort of "damping" effect of the tail fin is doing the job. This particular airplane is probably of the kind that "really" requires a little bottom rudder in the turn—for the reasons just explained. At the same time, the adverse yaw effect of holding aileron against the overbanking tendency should "really" require some top rudder. What we do is to leave the rudder more or less alone and let the adverse yaw effect cancel the tail fin's turn-slowing effect.

But there are airplanes that during a turn require positive unmistakable bottom rudder. What sort of airplane? As between airplanes of the same size and general shape, the slower one is more likely to be of that sort. Slowness makes the radius of turn, for given angle of bank, much shorter, and hence the curving path much curvier; this means a more pronounced sluing around of the tail and hence more "damping." As between two airplanes of the same speed, the bigger one is more likely to require bottom rudder in a turn, since, for a given curve, the fin that sits on the longer tail will be slued around more. As between two airplanes of same general size and same speed, the one with the longer tail will have the more powerful damping and hence will be more likely to require rudder in a turn.

This may clear up a lot of little puzzles.

It explains why many gliders are flown with very pronounced bottom rudder in turns. The position of stick and rudder in some gliders during a steep turn is shocking to the motor pilot; the rudder is well over to the inside of the turn, the stick is way over to the outside, because of the pronounced overbanking tendency of those ships; by all normal standards the controls seem badly crossed. But the airplane does not feel as if it were skidding, and, when the motor pilot tries some turns himself, he finally winds up with the same shocking disposition of stick and rudder. The answer is, of course, that the glider is a big aircraft and at the same time a very slow one; the turn-slowing effect of the tail fin is therefore very strong.*

This turn-slowing effect of the tail fin may also explain still another puzzle—why all the old-timers were so convinced that the rudder was the airplane's turning control. The ships of 20 years ago were no smaller than ours are, type for type, but very much slower; hence their turns, for the same angle of bank, were much tighter, their curving paths much curvier; hence the turn-damping effect of their tail fins was much more noticeable, and the need for constant inside rudder during a turn was probably much more common—thus leading to the fallacy that the rudder actually produced the turn.

But all this is strictly forbidden knowledge for the student. The student will achieve the best turn if he thinks of the rudder as a device to counteract aileron yaw and if he therefore relaxes on his rudder during those parts of the turn. For, in most airplanes, that is all the rudder does.

And it should be remembered that, whatever the disturbances are which the pilot must correct by rudder, they are still all sensed in the manner described. If the pilot feels himself pulled to the left, he must still give left rudder; if pulled to the right, right rudder—that goes even if the disturbance of the ship's flight should be caused by high-brow and rare effects such as the tail-fin effect just described. That effect has been described here not because it requires a different piloting technique but merely to clear up the confusion of some observant pilot who finds that what he actually does with his rudder in a turn is not entirely explained by the simple theory of the turn.

WHEN A TURN GOES SOUR

What should a pilot do when a turn goes sour? At first glance, it will seem that there can't be any one answer. It would seem that he should simply stop whatever fault has made his turn go sour—whatever that fault may be. If the nose is too high and the airplane losing

---

* Another effect is also important in gliders and must affect also some long-span airplanes—the tremendous span, coupled with slow speed, and hence a very curving turn makes for a great difference in speed as between the two wing tips during a turn. This not only produces a strong overbanking tendency, but it also produces an extra drag on the outer wing tip, which tends to straighten the ship from the turn, requiring an additional bit of inside rudder.

speed—let the pilot get the nose down. If the nose is too low and the airplane picking up speed—let him get the nose up. If there is skid, why, let him get off that rudder, and so forth. In short, it will seem that there will be as many different remedies as there are kinds of trouble.

But a closer analysis shows that this is not so. The pilot probably does not know what has made his turn go sour; if he did, he would have checked it in ample time, and the turn would never have gone bad! And an attempt to remedy the trouble direct—for instance to get the nose up or to get the nose down—will often fail to work, as we shall see. It is often like trying to treat the symptoms of the disease instead of curing the disease itself.

Only one thing will always work—stick forward. Leighton Collins's investigation of flying accidents suggests that this is probably the most important single rule in all flying; when anything goes wrong during a turn get the stick forward. *Don't* get your nose down. *Don't* get your nose up. *Don't* try to stop skid or slip; don't bother with anything, but get the stick forward.

"Forward" does not necessarily mean all the way forward, with positive forward pressure—but far enough forward to relax the back pressure. For, in the last analysis, it is the back pressure the pilot holds on the stick that causes his troubles. As long as an airplane flies at low Angle of Attack, with the stick near the neutral position, there simply is nothing that can happen to it, short of stupidly flying into a brick wall. You simply cannot lose control of it, nor will it lose control of itself. It will be stable, well behaved, safe. No gust effect can seriously harm it. No misapplication of stick or rudder can cause serious trouble—you may skid it all you want to. An airplane with the stick near neutral will always do whatever is necessary to maintain healthy flight. But get the stick back, and you become vulnerable. At high Angle of Attack the ailerons become tricky, the adverse yaw effect becomes treacherous; there is the possibility of the wing tips' being stalled by use of aileron, or misuse of the rudder may now stall one wing or the other and cause a spin. Any trouble is serious trouble when the stick is back. Thus the fact that the stick is back is the indirect cause of many troubles. It is also, however, a direct cause of trouble. It is the back position of the stick that loads the airplane with centrifugal force, so that it can stall although its speed is well above normal straight-flight stalling speed. It is the back position of the stick that causes the great stress on the wing's structure in a turn, due to the tremendous wing force developed. Get the stick forward and you are O.K.

"But getting the stick forward in a steep turn," someone may object, "doesn't put the nose *down*. With the airplane banked so far over on its side, what seems like down to the pilot is actually over, toward the horizon. Don't you want to point your nose *down* at the ground first of all?" But that is precisely where a pilot is likely to make his mistake. First of all, remember that what feels like "down" to the pilot means "down" to the wings. If the pilot is heavy, not toward the ground but toward the horizon, so is the airplane. In a steep turn the wings exert their lift not only against the force of gravity but also centrifugal force; hence, if you let the stick come forward and let the nose "dip" in the direction in which you are being pulled, that is out toward the horizon, you thereby relieve the airplane just as truly as you relieve it when you let the stick come forward in a wings-level ordinary stall.

In either case, what you really do is to reduce Angle of Attack. It is true that, in a steep turn, letting the stick come forward does not so immediately mean a gain of speed. The speed gain comes only through a series of other adjustments which the airplane has to make first. And, since the airplane stalls at a higher speed during the turn than during straight flight, once you let the stick come forward you will find that you really have enough speed anyway. And remember that the direct immediate cause of a stall is not "lack of speed" anyway, but simply excessive Angle of Attack. The moment you let the stick come forward, the ship's Angle of Attack reduces itself. For that's what the stick really is—an Angle of Attack control. Once your Angle of Attack is reduced, your stall disappears, control is again firm and safe, and your troubles are over.

"But if you let the stick come forward in a steep turn," someone will object, "you thereby stop the turn. The airplane will then just simply hang there, 'way over on its side, and a violent slip will result." This is perfectly true, but it does no harm, and it is necessary. You simply cannot stop your troubles without to some extent stopping the turn—because the turn is the cause of your troubles. It has been explained earlier that back pressure on the stick, tightness of turn, *g* load, nearness to the stall, are all really the same thing. Get rid of the back pressure, and you get rid of the turning; but you also get rid of all those other disadvantages. Your troubles are caused in the last analysis by excessive turning; stop the turning and you stop your troubles.

"But what about the resulting sideslip?" It is true that if you let the stick forward during a steep turn, you will suddenly hang there on your side, and slip off. But sideslip has never done anybody any harm. It is an extremely safe maneuver. Loss of control is just about impossible, and even if the pilot should fail to stop the sideslip the airplane's stability would stop it anyway; the dihedral will tend to level the wings, while the vertical tail area will tend to nose the airplane down and around, and the ship will turn into a steep fast spiraling glide. Moreover, once the back pressure is released, once the airplane is at low Angle of Attack, you can make unworried use of your ailerons to level your wings.

Just what should be the pilot's second step—after letting the stick come forward—depends of course on what the pilot wants. If a turn feels strange or unsafe, he generally wants to return to straight and level flight. In that case, relaxing all back pressure is the first step, and after that the airplane can be leveled and straightened. But a pilot might also choose to continue his turn. And it is perfectly possible in a steep turn to relax a good deal of one's back pressure, thus making sure the airplane is not stalled, and then resume part of back pressure, adjust the bank, and continue the turn. The appearance of the whole maneuver is simply a momentary hesitation of the nose in its travel along the horizon; the pilot momentarily feels sideslip; and then upon resumption of part of the back pressure the turn continues but in a safer more moderate form.

Note that it makes no difference whether you want merely to regain firmer control of a turn that somehow feels wrong or whether you want to stop the turn entirely and get back to straight flight, the thing to do first in either case is to let the stick come forward.

If any mere rule will ever make the difference between life and death, this is it. Let us therefore try to eradicate all doubts about it. Let us now investigate the various other remedies that a pilot might be tempted to use when a turn goes sour. We shall see that every other remedy is either dangerous or futile or both.

NOSE LOW?

Suppose the nose is low during a turn or that it suddenly falls below the horizon with a resultant loss of altitude, and probably an increase in speed. A pilot who is confused and has perhaps picked up the foolish notion that "in a steep turn the rudder becomes the elevator," may now try to bring the nose up by simply booting on top rudder. This is futile. The nose is down for a reason; probably the wing force is insufficient, either because the pilot holds insufficient back pressure on the stick or maybe because he holds too much and has stalled his wings. Just simply yawing the nose up does not remedy the trouble. At best, it is useless. The nose can be held up on the horizon by rudder, but the airplane will simply keep on losing altitude just the same in a slipping turn. At worst top rudder to keep the nose up is dangerous. If the nose is down because the airplane is stalled or almost stalled, then kicking it up may trigger off a "spin over the top," that is, the airplane suddenly levels its wings, rears up its nose, and, if the turn was a right-hand one, spins off to the left.

Perhaps the pilot analyses his trouble more correctly. Perhaps he recognizes that it is lack of wing force ("lift") that makes the nose ride low, and then he tries the apparently logical remedy—haul back harder on the stick.

This may work, but then again it may not. If the ship has been doing a diving turn and the speed is brisk, then application of enough back pressure to get the nose up may put excessive g load on the airplane. This has already been discussed earlier, where it was shown that such a maneuver is really a pull-up from a dive combined with a steep turn and means a doubled g load. This heavy g load may lead to structural failure, and it may also stall the airplane—despite its brisk speed. For remember, if the stick is back far enough, and thus the g load is great enough, the airplane will stall even at very high speed.

But the worst possibility is this: What if the nose is down because the airplane is already stalled? The airplane may have stalled during the turn without the pilot's clear knowledge. The thing to keep in mind in this whole discussion is that something has happened which the pilot did not expect. He didn't want the nose to be down, yet it is down. Hence the pilot's judgment is really not worth much in this situation. Whether he feels pretty sure he is not stalled is not material; he might be stalled. Such pilots as survive a spin-in frequently testify that they did not recognize the stall for a stall and did not even recognize the spin for a spin! Remember, in a steep turn, the stalling speed is much higher than in straight wings-level flight. Hence all the stall warnings concerned with speed—sound of speed, the feel of speed, the instrument indications of speed—are absent. Remember also that during a steep turn, when the stick is well back and the airplane nearly stalled, a gust may throw it temporarily into a stall. Actually, when the nose goes down during a turn, the chances are pretty high that it is because the airplane has been stalled.

But if the airplane is really stalled, putting more back pressure on the stick won't make the nose come up. It will only make the stall worse, and the nose will go down more. The pilot, still unsuspecting, will then probably haul back even harder; the stall will become worse, and the ship will spin off. Or perhaps the pilot then finally realizes that something is wrong and hastily tries to regain straight and level flight by unbanking, using his ailerons. Result—a stumble over the ailerons, and a spin. In short, once your nose is down during a steep turn, you may get it back up by hauling back on the stick; but chances are you won't. Stick forward will do the trick. Once your stick is no longer so far back, the airplane is again completely controllable, and you can do with it as you wish.

NOSE TOO HIGH?

Next, consider what a pilot might do if in a steep turn the nose is too far up, and the airplane probably losing speed rapidly. It is probably a toss-up. A pilot's first reaction might be to let the stick come forward—either because he understands what it is all about, or perhaps more naïvely because he still thinks of the stick as an up-and-down control and reasons, "Want to get the nose down? Get the stick forward." In any case, it is the correct reaction. If the nose rides too high during a turn, the thing to do is to relax some of the back pressure and let the stick come forward.

But in too many cases the pilot's first idea is to kick the nose down to the horizon by booting on bottom rudder. This may work in a rough sloppy fashion. But, if the airplane should be fairly near the stall, the sudden bottom rudder may spin it off—because the sudden yaw produced by the rudder throws one wing forward and the other wing back, thus stalling the latter while unstalling the former. There is irony in this; the pilot is conscious that he is losing speed, and he is afraid of stalling and spinning. He does what he thinks is the logical thing—points the nose at the ground. But, in the process of pointing it down, he produces the very trouble he was trying to avoid.

So you can see how the argument runs: when a turn has gone sour, it *may* be possible to remedy it by ailerons, rudder, increased back pressure or what not; but it is always only a chance, and there is always at least an equal chance of trouble. The only sure way to regain control is to relax the back pressure first of all. Once back pressure is relaxed, the airplane becomes again stable, controllable, and well behaved and the pilot can do just about whatever he wants.

Chapter 13

STRAIGHT AND LEVEL CRUISING

IT BORES the student pilot to fly the airplane straight and level at cruising speed; but connoisseurs, such as flight instructors and government inspectors, consider straight and level flight a most revealing test of a pilot's ability: it reveals whether he really understands how his controls work.

And, quite apart from its training value, straight and level cruising is important in practical flying. Most of your air time, after all, is going to be spent that way. If you do it efficiently, you will get cross country efficiently; for cross country hinges on straightness of flight. If you set out in the right direction and then fly straight, you can't help getting there! If you wander all over the landscape, you miss your landmarks, and you never know exactly where you are going or where you have been; and you are continually on the verge of getting lost. Also, if you fly straight, your compass behaves much better, even in rough air; if you wobble, the compass will swing, even in smooth air, and give you no accurate reading. Thus, the straighter you fly, the straighter you can fly!

Furthermore, the two nastiest drawbacks of flight, airsickness and excessive fatigue, have more to do with correct straight and level technique than most pilots realize. The small sideways skids that slosh the liquid around in your compass bowl and tilt up your compass card and make it swing—those same wobblings do quite similar things to your stomach and to your inner ear! And as for boredom, if you don't understand the technique of straight and level flight, a long trip through rough air can indeed produce a boredom that is almost a pain; but if you know how, flight through rough air becomes a game that keeps you alert and gives you a certain kinesthetic pleasure.

SOME BAD HABITS

If the air is smooth, flying straight and level is rather simple. Still there are some points that bear discussion. In trying to compensate for the so-called "torque," one easily develops the habit of over-compensating, holding too much right rudder and at the same time constant slight left aileron, thus flying continually with the left wing low, in a slight slideslip toward the left. This is simply a nervous habit. It is true that once in a while some ship will be wing-heavy and need constant slight aileron pressure merely to keep level. But that condition should be corrected on the ground by rerigging or retrimming the airplane; and, even if a pilot finds himself aloft in such a ship, there is never any excuse for flying with one wing low. The remedy is simple—keep your wings level by consciously checking the positions of your two wing tips relative to the horizon; and use only as much rudder as you then find necessary to keep the "torque" effect from swerving you off.

In "crabbing" flight, in a cross wind, there is a strong temptation to hold rudder "against the drift"; you see yourself sliding across the ground to the left, and a bit of right rudder crowds itself on without your intention. Students will almost invariable do this; but even pilots of long experience have been caught at it (so has the writer). The tendency to consider motion over the ground more important, more "real" than motion through the invisible air seems to be very deep-seated. The reasons why drift can't be stopped by rudder, can in fact not be stopped at all, have been discussed earlier in this book. The remedy is simple: Don't.

HOW TO FLY BY COMPASS

Flying a compass course is something that the student pilot will almost certainly do wrong the first time he tries it. He will try to maintain direction by constant reference to the lubber line of the compass; the moment the lubber line wanders off, he will make a corrective turn; in short, he will try to fly straight by compass. This cannot be done. Once an airplane is in a turn, however slight, the compass goes crazy and no longer shows correct directions. This is because in a turn (even a skidded turn) the compass (being freely suspended in its bowl) banks itself up one way or another; and, once it is banked, the vertical pull of the earth's magnetic north pole becomes effective and turns the compass card.

It will help a pilot to know just how the compass misindicates. While the ship's nose points in a northerly direction the compass fails to indicate a turn or even behaves as if the ship were making a right turn when it is actually making a left turn, and vice versa. While the ship's nose points in a southerly direction, the compass is oversensitive to turns; even a slight right turn will make it behave as if a pronounced right turn were going on; a slight left turn will make it behave as if a sharp left turn were being made. While the ship's nose points in an easterly or westerly direction the compass is not unfavorably affected by turns; it shows them approximately as they occur; but with the ship on those headings the compass goes crazy if you pull the nose up or shove it down, or if you make sudden changes in throttle setting; it will then show a turn even though no turn is going on! The compass behaves best while the ship is headed southeasterly or southwesterly. If you are ever caught above clouds without blind-flying instruments, and have to go on down through them, or in any other situation where you must use the compass as a blind-flying instrument, fly in a southeasterly or southwesterly direction.

With this possible exception you cannot fly an airplane straight by using the compass—because you cannot use the compass unless the airplane is flying straight! To fly a compass course, you must fly the airplane straight by watching the ground below, or a distant cloud, or the stars (or, if you are inside clouds, by watching a special gyroscopic blind-flying instrument). You must fly it straight even if you should have a strong feeling that you are flying in the wrong direction. Fly straight for perhaps 10 seconds, preferably more, and let your compass steady down; then only look at it, take a reading, fly straight another 10 seconds, take another reading; and *then*, if your direction turns out to be wrong, make a corrective turn.

Even that turn, though, must not be gauged by compass, but by the ground, the clouds, or the stars (or special blind-flying instruments). There is no point in trying to "kick" the compass around into the proper direction, because, once again, as soon as you are turning, and as long as you are turning, your compass is crazy, and its indications almost meaningless, except to an experienced instrument pilot who knows all its tricks.

ROUGH AIR

It is in rough air that straight flight becomes an art—and the interesting thing about it is that you can do it all wrong and never know it; you merely think the air is much rougher than it actually is, and you take much unnecessary punishment. This happened to the writer for many years, until an able Army flight instructor straightened him out.

Here, as elsewhere in the art of flying, the natural thing to do is the wrong thing; the proper reflexes must be carefully exercised through proper training. The natural thing is to use the controls independently to keep the airplane straight and level: when one wing drops, lift it by aileron. When the nose swerves off its proper direction, kick it back by rudder. This is quite wrong, and anybody who really understands his controls ought to see that it is quite wrong; but still, it needs pointing out.

The right thing to do is to use the controls in a coordinated manner all the time—even though at first that may seem tedious and unnecessarily laborious. What it really amounts to is that straight flight consists of a series of S turns, the turns being shallowed out so much that the S finally becomes a straight line.

This sounds abstruse. The best way to make it clear may be a close-up, slow-motion movie of a pilot who flies through rough air the faulty way, with a sound track added to put into words what he thinks. All right: The left wing drops; the pilot thinks, "That's what my ailerons are for," and picks the wing up with his ailerons. He does not use any rudder, for he thinks, "My nose is still pointing in the right direction; I don't need to turn; therefore, I need no rudder."

Note right here that his thinking contains two faults—one is that he forgets the adverse yaw and how the rudder must always accompany the ailerons and tidy up the mess they make; the other is that he thinks of his rudder as a turning control, forgetting that an airplane is turned not by rudder, but by leaning it into a bank and then lifting it around by back pressure on the stick.

The movie continues: As the pilot gives right aileron, the left wing comes up readily enough; but the same aileron which picks it up also yaws the airplane to the left. First thing the pilot knows, his wings are level, but his nose is off to the left. He fails, however, to connect cause and effect; instead of blaming the aileron yaw and using his controls with proper coordination next time, he thinks, "She's getting rough; now she has turned me off-course." Or worse still, he thinks, "Didn't pick that wing up fast enough; left her in the bank too long, and she started turning on me; next time, let's be a little snappier on that aileron." Being snappier is a polite word for being rougher; and it will only make the adverse yaw effect that much worse.

At any rate, he sees his nose off to the left of his proper heading; and pronto, he kicks it back where it belongs—by rudder. Perhaps he does this because he is ignorant and thinks of the rudder as the airplane's turning control; more likely though, his trouble is that he doesn't think of this bringing back of the nose as a proper turn. He thinks he is only "keeping the airplane straight." "That nose swerved off on me just now," he thinks, "but I wouldn't let it; I kicked it right back."

Note what his error is here; actually, that small correction is a real turn. The airplane is going in the wrong direction, and he makes it go in the correct direction. He is changing the direction of flight; and that makes it a real turn, though a small one. And no turn is ever flown correctly by kicking rudder.

The movie continues: As the pilot kicks right rudder, the nose swings back to its proper heading readily enough. But the airplane as a whole—the heavy mass of it, moving at 150 m.p.h.—keeps moving in the old (slightly wrong) direction for a few moments, skidding slightly sideways through the air; an airplane will always do so if an attempt is made to turn it by simply kicking rudder. This skid gives an unpleasant little shove to the pilot's stomach and inner ear. "Gosh," he thinks, "how I hate rough air." The skid, persisting for a second or two, now has an effect on the inherent stability of the airplane, or its tendency to resist sideslip or skid; it now makes the right wing go down.

The pilot again fails to connect cause and effect. All he sees is that now his right wing is down. "Boy," he thinks, "she is sure getting rough," and he goes to work and picks up the right wing by aileron. Again he does not see why he should use rudder, because at this moment the airplane is headed in the correct direction. Again the adverse yaw effect pulls him around, this time to the right; and so the whole tiresome cycle repeats itself.

The remedy is obvious: simply use the controls as they should be used. Don't use aileron without using rudder at the same time to counteract the adverse yaw effect.

Get rid of any remnant of the idea that rudder should be used only when it is intended to turn or to stop a turn; for proper flying in a conventional airplane, rudder must always be used when the ailerons are used, even if the ailerons are used only to counteract a gust.

Finally, don't try to make turns, however slight, by rudder only; make even the slightest change of direction by a regular turn, banking into the turn properly, coordinating rudder with your aileron, holding the bank until your desired heading has been reached, unbanking then, properly coordinating rudder and stick.

It will make that rough air much smoother!

HOW TO LEVEL OFF

To hold a contact altitude is easy enough in smooth air—if you can once get the airplane trimmed properly for level flight. There are several methods of leveling off for cruising flight after a climb; it makes little difference what method you use, but it is important to use some method, systematically; or else you'll fumble for a quarter hour.

Here are some methods: On reaching your intended cruising altitude (or slightly before), you press your nose down with your flipper until your air-speed indicator shows the desired cruising speed. Then, manipulating your flippers so as to maintain that speed, you adjust your throttle so that flight will be level; finally, you set your stabilizer control or trim-tab control so as to "take the force out of the stick," that is, so that you will not need to hold any pressure either way against the flipper.

Another way of doing the same job would be this: On reaching your altitude, you set your throttle for the desired cruising r.p.m. and then, by means of the elevator, hold your nose higher or lower as necessary to obtain level flight—meanwhile keeping the engine r.p.m. steady by readjusting as necessary. This way, you will finally arrive at a flipper position and an air speed which you can maintain while also maintaining level flight and also maintaining the desired r.p.m. Finally, you set your trim tab so as to take the force out of the stick.

Still another way is this: You climb beyond your intended cruising altitude by a couple of hundred feet; you set your throttle for approximate cruising power, and then you go down to your proper altitude in a shallow dive, thus quickly picking up your cruising air speed

This method is sometime used on fast clean heavy airplanes because such ships, if leveled off in the usual manner, will take several minutes to accelerate to cruising speed, keeping the pilot busy unnecessarily.

In any case, however, it is important to use method and to be decisive. The tendency always is not to level off decisively enough and to keep on climbing. This is part of a general tendency, common to most pilots, to "hang on" mentally to any flight condition in which one is or has been for a little while. The same mental quirk will make you reluctant after a long cross-country flight to slow your airplane up enough for a good approach and landing.

HOW TO HOLD YOUR ALTITUDE

To hold a constant altitude in rough air is difficult and demands close attention. Except at very low altitudes, you cannot see your losses and gains of altitude by looking at the ground. Hence it is most important to glance at the altimeter every few seconds (without, however, letting the attention freeze on it), catch the first small indication of any rise or sink, and make corrections promptly.

Promptness is so vital for two reasons: The earlier a rise or sink is caught, the smaller and less hurried will be the correction, hence, the smoother the flying. And with the altitude never varying much and corrections never violent, the problem of altimeter lag will be less bothersome. If you find suddenly yourself at 1,100 feet when you should be at 1,000 and then get rid of your excess altitude fast, your altimeter may lag behind the actual facts; it may still show 1,025 feet when you are back down at 1,000 feet; and, if you level off when it shows 1,000 feet, you will be too low: a few seconds later the altimeter will have caught up with the facts and show 975 feet. Thus altimeter lag, just like any lag in perception, will make you overcontrol and make your flying rough.

Just how should one gain or lose altitude to get back to a prescribed altitude? The accepted technique is to use the stick, raising or lowering the nose as necessary, and to handle the throttle so that the r.p.m. remain constant.

Thus, if an updraft carries you up, you must push your nose down. This, of course, causes your air speed to build up; remember, the "elevator" is really the speed control. Hence you keep your engine from racing by throttling back. If a downdraft carries you down, you must haul your nose up; and, as the airplane loses speed, you must keep your engine from slowing down by advancing the throttle.

This is the technique now officially taught. Perhaps it should not be. It does permit altitude-keeping with greater accuracy than any other technique; but perhaps superprecise altitude keeping is not so important as our present training methods make it appear. It is important in aerial mapping; in some phases of military flying, especially in traffic patterns around training fields; in some phases of instrument flying; and, arbitrarily, purely as a test of an abstract skill, in pilot flight tests. It is unimportant in most phases of flying.

And this technique of holding altitude has the serious disadvantage that it teaches a use of the controls, throttle, and stick, that is exactly contrary to their true functions and contrary also to the manner in which they are used in instrument flying. It helps to instill in the pilots mind the dangerous, the positively lethal notion that the way to make an airplane go us is to pull on the so-called "elevator." And it has an added disadvantage. When the air is rough, this method will often force the pilot into very fast flight—shallow diving merely to keep from being carried up by an updraft; but, in rough air, excessive speeds make for excessive gust loads that may imperil the structure.

A sounder procedure for level flight would seem to be this: The pilot holds his stick—and hence his air speed—steady, and he regulates his altitude by retarding or advancing his throttle. For, in the long run, only one thing will ever make an airplane go up—throttle, that is, power. It must be admitted that this method is slower in making altitude corrections, but it uses the controls in a more proper manner and teaches the student sounder reactions.

In practical flying, the best answer to the altitude-holding problem is often: Don't hold your altitude. On a day with unstable air, the smoothest and most efficient cruising will result if you simply let the next downdraft undo what the last updraft did, merely holding an air speed throttle setting which in the long run keeps you at the same average altitude.

FROZEN ATTENTION

Be that as it may; in precision maneuvers, such as are required for license flight tests, constant altitudes must be kept. To keep them, it is once more most important to catch the first small direction; and to do that, you have to keep watching your altimeter; and concerning that, here is a trick.

Don't let your attention freeze on it, for if you did that, some other phase of your flying would go wrong. And at the same time don't let your attention freeze on anything else (such as your drift, for instance) for if you did that, your altimeter would go wrong. It is perhaps the most important mental characteristic of the good pilot that his attention is always dilated, spread, wandering—never concentrated. Maybe that is why it used to be thought that really good pilots were on the dumb side, or anyway not too brainy, and that too brainy a man made a poor pilot; the brainy, well-schooled man has a habit of concentrating on one thing at a time to the exclusion of all others, and of keeping concentrated on it until it is solved or finished or done.

That you can't do with altimeters. An Army flight instructor taught the writer a trick that has been useful ever since: Don't ever concentrate on your altimeter long enough to read it, he said. Just glance at it, and then look out again at the ground, horizon, and so on. Your eye will still retain the image of the dial, and you can then read it at leisure while looking at something else!

In this connection, it seems advisable for a student to study the altimeter face closely and actually practice reading his altimeter on the ground. Have you ever noticed how often one's first instinctive interpretation of the altimeter is exactly wrong? You see the hand at 300 when it should be on 400, and your first reaction is that you are too high; and that sort of thing. For the family airplane of the future, someone ought to design an altimeter with a face similar to an ordinary household thermometer. He would help the owners to do better flying and would help in overcoming the awe in which the average man holds an airplane's instrument board.

Still on the subject of attention freezing, and also on the subject of "hanging on" mentally to the flight condition in which one once is: You certainly can't afford, in a flight test, to watch your altimeter while it returns from an off reading to its proper reading. The tendency to do so is very strong. All you can afford to do is to give a quick glance, take notice that you are too low, put on some back pressure and more throttle, and keep your attention wandering where it ought to wander—flight path, coordination, planning, and so on. Then, after a minute or so, the airplane is bound to be back at proper altitude, and that is where you're likely to make a mistake.

You see your altimeter showing proper altitude, and you forget what you did a minute ago: you forget that the airplane is in a slightly climbing condition. The airplane does not look as if it were climbing, and it is just then at the proper altitude, and you are naturally reluctant now to reduce throttle and put the nose down a bit. But if you don't thus take off the correction you put on a minute ago, you're going to be above your proper altitude in another minute!

Such are the headaches of precision flying.

Chapter 14

THE GLIDE

THE glide is the maneuver that first shows the student that flying may not be quite so simple after all, that the airplane has a perverse way of not going where you are trying to make it go. The other basic maneuvers—the climb, the turn, straight and level flight—are also extremely hard to do accurately. But in those maneuvers we are usually well up in the air, and the pitiful drunklike unsteadiness of our control does not show up so clearly. A mile up, a couple of hundred feet more or less altitude makes no noticeable difference; the altimeter needle points to a different figure on the dial, and that's all. And a mile up, you may be half a mile to the right or left of your intended position—and you will hardly notice it unless you look sharp. But the glide takes you back, out of those mile-wide spaces of the air, down toward the narrow, restricted world of the ground—until finally the airplane must arrive, within a few feet exactly, on the intended spot on the runway. And when the student first tries to do it, he overshoots or undershoots his aimed-for spot by thousands of feet!

The difficulties of the glide are twofold. One is the problem of perception: the inexperienced eye simply cannot gauge exactly enough where the airplane is going. The other difficulty is a problem of control: in the glide, the airplane's response to the stick is almost directly contrary to "common sense." Sooner or later, as the glide progresses toward the ground, even the inexperienced pilot's eye perceives that he is going to miss his spot, and he will then "naturally" want to take corrective action. If he is overshooting, he will "naturally" point his airplane down more steeply; but this "natural" reaction will only make him overshoot still more. Again, if he is undershooting, he will "instinctively" want to point his nose down less steeply and try to "stretch" his glide. But again, this instinctive reaction will only make him fall even shorter and may actually cause him to stall, spin, and crash.

The problem of perception in the glide will be discussed in another chapter of this book. The present chapter deals with the problem of control—what the pilot must do to make the airplane go down more steeply or less steeply—and, equally important, what he must not do.

Curiously enough, this is at the present considered forbidden knowledge. It is not usually taught to students. Flight instructors believe that a student should make no attempt to steepen or shallow his glide path, that he should always use the same glide, that is, the same gliding speed, the same gliding attitude of the ship, the same steepness of glide. The burden of getting the ship down onto "the spot" is to be entirely upon the student's perception; instead of "playing" his glide, he should "play" his turns and his patterns; he must correctly choose the spot and the moment to close his throttle and begin his glide. He must then maneuver intelligently along one of the various standardized approach patterns; and he must properly time his final turn toward the intended landing spot—all the time holding that constant "normal" glide.

There are good reasons for this practice. The problem of visual judgment of the glide path is so difficult that the student simply can't master it if each time he tries an approach he uses a different glide, of different speed and different steepness, or if he varies the steepness of his glide in the midst of any one approach; to do so would introduce too many variables into a problem that is already intricate enough, and it would leave him confused. Furthermore, control of the glide path sometimes requires that the pilot slow the airplane up, possibly that he fly it quite near the stall; and instructors don't care to contemplate slow flying by students whose sensing of speed and "lift" has not yet become subconscious. Especially near the ground, when the student is tense anyway and preoccupied with hitting his spot, and when he thus has not enough attention to spare for his airplane's speed and "lift," such flying is too likely to result in stalls, spins, and crashes.

These objections, however, are all concerned purely with training —with the business of *learning* to fly rather than of flying. Experienced pilots do actually play their glides heavily. In fact, the trick of glide control is among the most important tricks of the pilot's trade. And there are good reasons why even the student pilot should at least know and understand how the glide is controlled, why he should even be allowed some supervised experimentation with glide control. For, as has been pointed out, when the student notices that he is overshooting or undershooting, he is bound to make some attempt to steepen or shallow his glide, whether the instructor approves of it or not. In case of undershooting in an actual emergency landing, his attempt to stretch his glide will be desperate indeed. His natural common-sense attempts at glide control, however, are exactly wrong and in the case of an undershot emergency landing will cause him to stall and spin. Hence, by being shown how steepness of the glide can be controlled, he will not only know what to do; he will *ipso facto* know also what *not* to do; and, in flying, knowing what not to do is often the more important part of wisdom.

THE SECRET OF GLIDE CONTROL

The secret of glide control is quickly stated: If you want to go down more steeply, point your nose down less steeply. If you want to go down less steeply, point your nose down more steeply. That's all there is to it! It is true, of course, only within limits, only of maneuvers that can properly be called a glide, rather than outright dives, or outright stalls. The nose is lowered or raised only by a few inches; but those few inches are surprisingly effective. The rule just given runs contrary, of course, to all common sense—contrary to all one's experience with automobiles, sleds, boats, and other ground conveyances. And, because it runs contrary to one's experience, it is extremely hard to do the right thing—especially in an emergency when glide control really matters; and it is extremely hard to refrain from doing the wrong thing. But actually it is entirely logical and quite in accordance with common sense. If you can see the common sense of it clearly enough, it won't be nearly so hard to do.

Let us start with a simple case—a case so simple that it is not quite realistic. The realism will be filled in later. Let us assume that the day is completely calm. A student is flying an airplane that cruises at 100 m.p.h. and stalls at 40 m.p.h. The student makes all his glides at exactly 75 m.p.h. For that is the speed which the instructor has demonstrated to him as the "normal" glide of that particular airplane. He knows exactly how the airplane sounds at that speed, how his controls feel, how much its nose is pointed down, and so forth.

Those clues to speed and "lift" have been discussed in an earlier chapter of this book. And he also knows approximately how steep this ship's normal glide is. Let us assume that it is 1:10; for each foot of altitude, the airplane covers 10 feet of distance.

Our first job now is to analyze this "normal glide." Just why has the instructor fixed this particular glide as "normal"? Why does he not recommend a glide at 90 m.p.h. or one at 60 m.p.h.?

The instructor has set up this glide as "normal" largely because it is the best compromise between two evils. If the student glided much faster, he would arrive at ground level with too much excess speed which he would have to dissipate before he could make a three-

The secret of glide control: if you want to come down more steeply, point the nose down less steeply. If you want to come down less steeply, point the nose down more steeply.

point landing; for the three-point landing is made at stalling speed, or close to it. The dissipating of all that speed would mean a long-drawn-out, "floating" landing, and this "float" is evil number one. Evil number two is loss of control or an actual stall. If the student glided his ship much more slowly than the "normal" glide, he might inadvertently stall himself—especially since his attention is on the ground rather than on the ship. Also, as has been pointed out in the chapter on The Ailerons, the controls of an airplane become somewhat inefficient and tricky in slow flight even if the ship is not actually stalled. Thus the normal glide is a compromise between dangerously slow speed and awkwardly excessive speed.

THE NORMAL GLIDE

Because it is a compromise between excessive speed and excessive slowness, the "normal" glide is also the airplane's most efficient glide.

It is the glide that results in the shallowest path of descent: in which the airplane will cover, from a given altitude, the longest distance. *That* this is so, and *why* it is so is the core of the whole business of glide control. It should be clearly understood. If you want to glide an airplane very fast, you must point its nose down very steeply—since at high speed the fuselage, landing gear, motor, and propeller develop terrific drag. Hence a very high speed glide results in a steep glide path. The airplane in our example, which glides 10:1 at 75 m.p.h. would glide only perhaps 6:1 at 100 m.p.h. In fact, pilot's language would not call such a maneuver a glide at all, but a dive.

If, on the other hand, you glide an airplane tco slowly you also get a steep path of descent. Pilots would say that this is because in slow flight the airplane mushes excessively. Engineers would say it is because in slow flight the airplane proceeds at high Angle of Attack and is being braked by much induced drag.

The two expressions mean the same thing. If the airplane of our example were glided at 50 m.p.h., its nose would point about level, approximately at the horizon, much as it points in ordinary cruising flight. But it would not go where its nose points but would instead sink ("mush") downward continually; the angle between where it points and where it goes being the Angle of Attack. Its actual flight path would be steep. Hence a very slow glide also results in a steep descent.

The "normal" glide, then, is a compromise between too-slow mushy glide and the too-fast divelike glide. In airplanes of average design, this best compromise between "dive" and "mush" occurs at a speed about halfway between stalling speed and cruising speed. In airplanes that are exceptionally clean, it lies a little more on the fast side, closer to the cruising speed. In airplanes that are encumbered with lots of struts, wires, open cockpits, and uncowled engines, the normal glide comes more on the slow side, a little closer toward stalling speed. But always the "normal" glide is the glide of shallowest descent; the glide at which (in still air) the airplane will cover the greatest horizontal distance from a given altitude.

That, then, is the condition of the student who approaches his intended landing spot in a "normal glide." He is gliding the airplane in the most efficient maneuver possible and is getting the shallowest possible glide path. And now, suppose he discovers that he is undershooting. What can he do?

UNDERSHOOTING

What he *wants* to do is obvious. He wants to raise his nose a little, holding the stick farther back, and simply point the airplane not so steeply down. And that is, of course, exactly the wrong thing to do. By doing it he slows the airplane up, and, as it slows up, it goes to higher Angle of Attack: it begins to mush. The glide path steepens, and the maneuver accomplishes the contrary of what was intended!

And that isn't all that's wrong with it. The worst part of it is that the steepening of the glide path does not follow immediately. The *first* reaction of the ship is deceptive—there is a *temporary* ballooning which for a couple of seconds markedly shallows the glide path. It's only after those few seconds, when the airplane has slowed up, that the mushing and the steeper descent begins. This first brief reaction of the airplane tends to trick the pilot into thinking he has done the right thing when he has actually done the wrong thing. He was too low, to begin with, and instinctively brought his stick back. The airplane balloons, moving forward without losing much altitude for the moment, and the pilot thinks proudly, "Well, I fixed it." Then the mushing gradually develops, and still a few moments later the pilot discovers that he is still too low, still undershooting. He is too low, too short of the field, largely because he raised his nose, slowed the airplane up, and caused it to mush; but he thinks he is too low *although* he raised his nose! "I did the right thing" he thinks, "but apparently I didn't do enough of it." And so he brings the stick back another bit and holds his nose higher by another bit, and again the ship's *first* reaction is a brief shallowing out of the glide path, which again hides from him the fact that he is really all the time making his glide less efficient. First thing he knows, he has not only undershot the field by quite an unnecessary margin, but he also finds himself near the ground, all slowed up and near the stall, slow and low and thus in trouble.

Well, what else could he do to stretch his glide? In our present, somewhat artificial example, he can't do anything at all. The normal glide is (by definition) the glide in which the airplane will have its shallowest descent. Therefore it makes no difference whether the pilot pulls his nose up more or points it down more—he will only steepen his descent. In the normal glide the airplane is doing its best.

Put it into any other sort of glide, and it will necessarily do a little worse.

From this follows the most time-honored rule of flying: *"You can't stretch your glide."* As we shall see later, this rule must not be taken quite literally. In practical flying, most pilots don't use the so-called "normal" glide; also, there is usually a wind, and wind has an important effect on the whole problem. Hence there are tricks by which you can stretch your glide. But the rule is certainly true as regards any naïve, "instinctive" attempt to stretch one's glide—any attempt to get a little farther simply by holding the airplane's nose a little higher.

OVERSHOOTING

Now suppose that the student, descending in a normal glide, discovers that he is overshooting his spot. What can he do to steepen his glide?

At first glance, it would seem as if he could do either, pull his nose up or point his nose down. It has just been shown that *any* change from the normal glide will steepen the path of descent. The student's natural tendency of course will be to point his nose down in order to get down more steeply. This will do the trick, but only under certain conditions.

It will work if the glide is a very long one. If you are at, say, 5,000 feet and want to get down steeply, a steep, very fast glide, or a shallow dive will do the trick. But the whole problem of glide control arises only in the last few hundred feet of altitude, just before landing; and in such a short distance and time, putting the nose down will *not* bring you down shorter. Dropping the nose will result in a steeper glide path but also an increase in speed, and hardly has the speed increased when it must be got rid of again so that the landing can be made. The flare-out then carries the airplane forward just about as far as it would have got if there had not been any attempt at correction, but a normal glide had been maintained all the way down.

It may even carry you a little farther! This is because this flare-out, this long "float," will be made close to the ground; and close to the ground an airplane is quite markedly more efficient than it is a little higher up in the free air, owing to the cushioning effect of the ground which catches the downwash from the wings and thus indirectly gives the wings more lift with less "induced" drag.

No—if you are overshooting, pointing the nose down won't help. The only remedy is to pull your nose *up* and get the airplane into a mushing glide. Your instructor won't like it, and it is nothing to play with if you can't trust your perception of speed and "lift" on that particular airplane. But it does work.

SHORT RUN AND LONG RUN

It works, however, only after a certain delay. The airplane's first response, lasting several seconds, is a ballooning as already described; the mushing and the steeper descent start only after the airplane has slowed up. This first ballooning can be quite disconcerting. You pulled your nose up because you were too high, because you wanted a steeper descent; and next thing you know you are shooting forward almost without losing any altitude! But if you just hold everything for a couple of seconds, you will get your steeper descent.

And not only will your descent be steeper, but you have then the additional advantage that you will arrive at ground level in slow mushing flight quite close to the condition in which you want to land. Thus there will be no prolonged floating, and not only will the approach be nice and steep, but the landing itself will be nice and short.

But remember, slow glides for a student are like fire for a child—he has to learn to play with it, but he has to learn with caution. A slow mushy glide is a condition close to the stall. In such a condition you cannot afford to make bad, skidded turns. You cannot afford to tighten up on the stick or to let the speed and buoyancy of your ship temporarily fade from your attention. You have to react correctly. If you should suddenly find yourself stalled, or if the bottom should drop out from under you, or if one wing should drop, you cannot afford to come back on the stick or to attempt to level the wings hastily by sudden hard aileron; if anything goes wrong, feels wrong, or sounds wrong, your first reaction must be to get the stick forward, thus reducing Angle of Attack and regaining control. You must trust yourself to do this even if you should be close to the ground. And that's a large order.

Just how much an airplane can be slowed up with impunity to get a steep descent depends upon air conditions; in rough air, any airplane needs more speed. But it depends even more on the airplane's stalling characteristics. Some are so nasty that you would never slow them up much below the normal glide. A few are so well behaved that cocky pilots sometimes bring them down actually stalled, shaking and buffeting and with the stick almost all the way back. In such a very slow approach the path of descent is extremely steep. But the airplane is then in no condition to check its descent by the landing flare-out—there simply isn't enough reserve "buoyancy." And if it were allowed to glide into the ground in that condition it would flatten its landing gear and crack up. Hence pilots of such ships develop the technique of "dumping." Just before landing, at perhaps 50 feet of altitude, the stick is let forward, and the nose is allowed to drop. The airplane then picks up new speed at the same time, of course, getting a sudden sharp sink. Hence the dumping must be done early enough to avoid hitting the ground. Once the airplane has that new additional speed, it is then in a condition close to the normal glide, and a normal flare-out can be made to check the descent, followed by a normal landing.

WHAT PILOTS ACTUALLY DO

But now let us make the story more realistic. In actual practice experienced pilots don't use a "normal" glide during their approach, but a much slower glide. Most pilots actually slow the airplane up as much as is compatible with safety; they glide it just fast enough to retain good control. On the ship in our example, the experienced pilot would make his approach probably not at 75 m.p.h. but more nearly at 65 m.p.h.—in smooth air even at 60 m.p.h. He would do so for various reasons. In the first place, by bringing the ship in more slowly he gets it to the ground in a better condition for an immediate landing; he won't have so much speed to get rid of or be bothered by so much tendency to "float." In the second place, the slower mushier glide gives him a steeper descent, and the steeper descent is much easier to judge. There is no particular reason to glide an airplane at its most efficient speed—the flatness of the glide is a liability, not an asset. Some extremely clean airplanes have so shallow a "normal" glide that an approach in that condition is just about impossible to judge and may be actually dangerous because the ship would have to be brought close across obstructions in the vicinity of the airport.

HOW TO STRETCH YOUR GLIDE

But the most important advantage of the slower approach glide is that it gives you a chance to stretch your glide.

Consider: You have an airplane whose "normal" most efficient glide, resulting in the shallowest path of descent, is 75 m.p.h. You fly that airplane at 65 m.p.h. Any time then that you are afraid of undershooting, you merely need to drop your nose an inch or two, and the airplane will presently become more efficient and its descent shallow out.

The pitfall of glide control: when doing the right thing, you get the wrong result first, the right result only later. Nosing the airplane up steepens the descent, but *first* causes a temporary ballooning. Nosing the airplane down shallows the descent, but *first* causes an extra sink. Be patient for a few seconds.

That's why it is true that in practice you can stretch your glide. That's why one can put it as a rule—one of the most important rules in flying—that in the glide, if you want to get down more steeply you must point your nose down less steeply. And if you want to get down less steeply you must point your nose down more steeply.

In practice, it goes against one's grain to push the stick forward when one is too low anyway. The student may find it a little easier if he remembers that our controls are wrongly labeled. The "elevator" is not the airplane's up-and-down control, but its Angle of Attack control, or, if you will, its speed control. Thus the apparent nonsense of getting the stick forward in order to shallow out one's descent turns out to be pretty good sense: you simply put your speed control in the (imaginary) notch that will result in the most efficient gliding speed.

Another thing that will trouble the beginner is that the ship's first, temporary response is here again exactly opposite to its long-run, real response. It has been explained above that when you raise your nose in the glide to steepen your descent, the first result is a ballooning, a temporary shallowing of the descent; and that descent steepens only after this ballooning effect has spent itself. The same thing happens, with the signs reversed, when you drop your nose in the glide in order to make your descent more shallow. The *first* result of dropping your nose is a decided steepening of the descent, a quite noticeable sudden sink. This is likely to disconcert the pilot who is anxious about undershooting. He may conclude that he has done the wrong thing and hastily raise his nose again. But if he will only sit tight for a couple of seconds, he will presently hear his speed picking up and will presently see his glide path shallow out. A few seconds later, he will have made up for that sudden sink; a few seconds still later, he will be far ahead of where he would be if he had continued his original glide.

WIND EFFECTS

And now, let us add another touch of realism—the effects of the wind on this whole business of glide control. There is almost always some wind. The last, most critical leg of the approach is of course always made into the wind, since the landing itself is made into the wind. And this head wind is an important aid in glide control.

Take an extreme case first: Suppose that, in the airplane of our example (which stalls at 40 m.p.h. and has its normal glide at 75 m.p.h.), an approach is made into a wind that blows at 40 m.p.h. Obviously, by pulling his nose up and slowing it up to 40 m.p.h. the pilot could get a glide path of infinite steepness; the ship would sink straight down, making no forward headway at all. This glide-steepening effect would be quite independent of the mushing that would of course accompany so slow a glide. It would be simply a wind-drift effect. The ship would drift back with the air exactly as fast as it was moving forward through the air, and no net forward motion would result. This is an exaggerated example; the general fact is that, if you are gliding against the wind, you can steepen your descent enormously by slowing yourself up—even quite apart from any mushing.

Take another exaggerated example: Suppose the wind is 75

m.p.h. Obviously, the airplane of our example will then make no headway at all if flown at its most efficient glide of 75 m.p.h.; in that glide it will simply descend vertically down, owing to wind drift. Although it goes through the air in an efficient manner, its flight path relative to the ground is not efficient.

But let the pilot of that ship dive it, at, say, 100 m.p.h. and he will shallow out his descent; he will no longer get only down but will move forward as well. He could probably point his nose down 30 degrees and pick up speed well above his normal cruising speed and still make better distance than by maintaining his normal glide. His glide through the air is then much less efficient, because it is too fast, and the ship is nosed down too much. But his path relative to the ground will be more efficient. This, too, is an exaggerated example. The general fact is that, if you are gliding against the wind, you can usually shallow your path of descent by gliding a little faster than your normal glide, even though in still air such a glide would be inefficient.

There would be no advantage in giving exact figures here, since too much depends on the velocity of the wind and the speed of your ship. But, on a day with average wind, you will find that you can gain distance very effectively by pointing your nose down an extra bit and flying perhaps 5 m.p.h. faster than your ship's "normal" glide; and, on a really windy day, it may actually pay to make the approach in a shallow dive. Inversely, a slowing up of the glide by 5 m.p.h. is doubly effective if there is a wind—it not only produces mushing, but it gives the wind more chance to set you back and thus to steepen your descent still further.

Here for once we have a case where there is no "on the one hand— on the other hand." *All* considerations, both of gliding efficiency and of wind drift, show that the way to steepen your descent is to point your nose down less steeply; by doing so, you diminish the efficiency of your glide, and you allow the wind to set you back. The way to stretch your glide is to point your nose down more steeply. By doing so, you increase the efficiency of your glide, and you allow the wind less time to set you back. Try it, and you will see that it really works.

YOU ARE BETTER OFF WITH YOUR NOSE DOWN

Of the two tricks—steepening the glide by pointing your nose higher, and shallowing your glide by pointing it lower, the latter is the more important. For there are other ways of killing altitude—for instance, a sideslip, or an S turn; but there is no other means of stretching a glide. But of the two tricks nosing down is also the more difficult to do, for psychological reasons. The natural tendency is to pull back on the stick and to point your nose away from the ground; it rubs your nerves the wrong way to put stick forward and point the nose down at the ground—especially when you are really low and in a real emergency. A pilot who has succeeded in recovering from a stall or a spin at low altitude will usually report that getting the stick forward at that moment was the hardest thing he ever did in his life.

Hence, here is a little sales talk, designed to help you do the right thing in a glide when the going gets tough psychologically, when the temptation to pull the nose up becomes too strong.

The thing to keep in mind is this: Even if your glide-stretching attempt should fail, even if you should not make it into your intended landing field, you will still be much better off if you have been doing the right thing and have kept your nose *down* than you would be if you had been doing the wrong thing and had tried to stretch your glide by holding the nose *up*. You will reach the ground in a better condition. Here is why.

Practically the only way to kill yourself in an airplane is to hit the ground nose first. It makes little difference whether the airplane be big or small, fast or slow—even a light plane, if it hits the ground nose first, is bound to kill its pilot. And even an air liner, if it makes contact with the ground in a glancing fashion, skidding on, is likely to give its occupants no more than a shaking up. This has been proved time and again, in all sorts of ships, coming down on all sorts of terrain, including forests, swamps, and urban back yards. If the ship has only a few seconds after ground contact in which to decelerate, sliding to a stop, the jolt becomes bearable for the human body.

Except for some freak accidents, practically the only way to hit the ground nose-first is to stall in or spin in. Hence the big problem in a rough emergency landing is to avoid a stall or spin and to get the airplane down to the ground under control. It must have enough speed and "lift" so that a proper flare-out can be made. Contact with the ground must be similar to a normal landing—preferably a rather fast "wheel" landing. The old-timers, whose engines let them down often, used to say that, if you had to land on rough terrain, the thing to do was simply to imagine a nice airport under you, and go ahead and land on it.

Consider now what happens to the fellow who is convinced that pulling the nose up will stretch his glide. The closer the ground comes, the more urgently will he try to stretch his glide and hence the higher he will hold his nose. Thus he is bound to approach the ground in very slow flight, quite close to the stall. Now, when the ground finally comes quite close and contact is imminent, any pilot (whatever ideas he may have about glide control) is likely to come back on his stick a little. This is simply an instinctive reaction that is almost as uncontrollable as ducking or blinking when something is thrown at you. On a fast-gliding airplane this reaction does no harm—in fact it is useful because it comes naturally just when it is time for the landing flare out. But, if the airplane is gliding near the stall anyway, this same "ducking" reaction is likely to bring on a stall, and a nose-first impact.

But that isn't all. Near the ground, there is likely to be some turbulence. Normally we get near the ground in slow flight only on the down-wind side of a large landing field where the air is comparatively smooth; hence most of us have not the proper respect for the roughness of air low over ordinary terrain. This turbulence only shakes up a fast-gliding airplane. But it can throw a slow-gliding airplane into a stall or can so disturb it that the pilot, reacting hastily on ailerons and flippers, throws it into the stall. "Fast-gliding" and "slow-gliding" does not mean, of course, any particular speed in terms of miles per hour. It means that the airplane is gliding fast or slowly relative to its own stalling speed. For instance, 90 m.p.h. would be extremely fast gliding for a light plane that stalls at 35 m.p.h. It would be rather slow gliding for an air liner which stalls at 65 m.p.h. To express it more precisely: fast-gliding means gliding at small Angle of Attack; slow-gliding means gliding at large Angle of Attack.

And even that is not all. Coming down on rough terrain, the pilot is bound to make a last-minute swerve to avoid some obstruction; and he is likely to make this turn a skidding one, misusing his rudder. Altogether, then, the chances of successfully "mushing" into a really rough forced landing are very small; the "mushing" is almost certain to develop into a stall or a spin. The old-timers used to say that, if your nose is up when you hit, you can walk away from it. But that doesn't mean that the approach to a landing could be made nose-high. If the approach is made nose-high, the actual ground contact is too likely to be nose-down. If you want to be able to get your nose up just before you hit, you must have kept it down until then.

Now consider the fellow who is thoroughly convinced that the way to stretch his glide is to put his nose down. Quite probably this technique will keep him out of trouble in the first place, because he may succeed in reaching that open field. But, even if he doesn't reach it, he is still better off.

The more anxious he becomes about the relentless approach of the ground, the more firmly will he keep his nose down; or at least, his firm intention to put his nose down will serve to counteract that instinctive "ducking" tendency to pull his nose up. When he finally arrives at treetop level, he has still plenty of speed. If now a fence or a pole line looms up between him and an open spot, he can use his excess speed and buoyancy to execute a last-moment zooming jump across it. Or, if he wants to make a last-minute swerve to avoid, say, a tree, he can safely make it. Turbulence can't cause him trouble. And, finally, he has the speed and the buoyancy with which to flare out, check his descent, and get his nose slightly up for the actual landing. Thus it is a mixture of psychological and physical effects that keeps him out of trouble in the first place and, should he be in trouble, protects him.

Glide control, in the exact form in which it is described here, does not work equally well on all types of airplanes. It works best on low-powered and medium-powered cabin ships of the kind most widely used in commercial or private operation and for training. Such an airplane is clean. Because of its cleanness it can fly on a small power plant and hence is economical to own and fly. Its glide is shallow because of its cleanness and also because its small motor, when wind-milling in the glide, does not have much braking effect. The "normal" most efficient glide of such a ship comes at comparatively high speed well toward cruising. On the old type of open-cockpit, wire-braced, powerful biplane—the sort that is still used in some places for primary training—glide stretching by dropping the nose does not work so well. On powerful, heavily wing-loaded, well-flapped airplanes the whole problem is altogether different—as we shall see.

In the long run most pilots will do most of their flying on clean low-powered small airplanes, and most of our flight training is on such ships. But, because open-cockpit biplanes have traditionally been used for training, many of our training manuals are still written around those airplanes. And, because the student will advance (or at least dreams of advancing) to big powerful heavily flapped airplanes, perhaps multiengined ones, some of the training routine is shaped toward flying such airplanes. But in the process, we sometimes forget to tell the pilot how to fly properly the ship he is actually flying.

FLAPS

In big heavily wing-loaded airplanes, equipped with flaps, the things that have been explained are fundamentally just as true; but they are overshadowed by other, more important effects.

The glide of a heavily wing-loaded airplane is much steeper than that of a lightly wing-loaded one. Airport opinion has it that an airplane glides steeply simply because it is heavy, simply because it has not enough wing surface. This is not correct. A heavily wing-loaded glider, for example, has no steeper glide than a light one. The heavily wing-loaded airplane glides so steeply because of the drag of its power plant when it is windmilling in the glide. A fast, heavily wing-loaded airplane must necessarily have a big motor and big propellers—big not only absolutely but also relatively—in proportion to the ship's weight. In engineering language, if a ship has a high wing loading, it must have a lower power loading. In ordinary language, if you have small wings, you need a big engine. But in gliding flight the power plant isn't pulling the airplane; it is being pushed by the airplane! The power plant acts as a brake. In a lightly wing-loaded ship, where the power plant is small in proportion to the ship's weight, that braking effect is not important. In a heavily wing-loaded airplane where the power plant is big in proportion to the ship's weight, this braking effect makes the glide quite steep.

Thus steepness of the glide makes the whole problem of glide control much simpler for the pilot of a heavily wing-loaded ship: the steeper glide is much easier to judge. Thus the need for any corrective action is less likely to arise. Furthermore, if you are a little too high, it is usually enough simply to point the nose down more and dive the ship a little; against the drag of the big power plant it can't pick up much excess speed.

In addition, most big heavily wing-loaded airplanes are equipped with flaps. Flaps have a double purpose of lowering the airplane's stalling speed and of increasing its drag, thus steepening its glide. As far as glide control is concerned, only this drag effect is important. With the flaps down, the airplane's glide is so exceedingly steep that there is hardly any problem of glide control; you simply make your approach high, close to the field, so that you can't possibly under-shoot; then you put on your flaps and are virtually certain not to overshoot. If you should be about to overshoot, you simply nose down more steeply. Against the drag of the flap, the ship can't pick up very much additional speed in the first place even if nosed down quite sharply.

Flaps have the additional advantage that any excess speed which you may bring down to the ground can't bother you much. As you flare out, the flaps kill your speed in a matter of seconds, and no long floating is possible.

All this in combination makes glide control on a heavy powerful airplane rather simple and quite common sense; if you want to get down more steeply, nose down more steeply.

THE POWER DESCENT

The glide is not the most practical way to make an airplane descend. It has many disadvantages. First and most important, accurate control of the flight path is too difficult, and even the most expert pilot, using all the tricks of the trade, cannot always steepen or shallow out his glide path enough to meet the requirements of the practical situation. Second—a fast airplane, descending in a normal glide, loses altitude so rapidly that the pressure change may hurt the passengers' ears. Third, the engine cools rapidly in a glide, and this may cause warping of valves and generally much engine wear, and may also foul up the engine so that it fails to "take" again when the pilot wants to use power again for some reason. Fourth—in heavily wing-loaded airplanes, especially with flaps extended, the power-off glide is so steep that the job of "flaring out" for the landing becomes rather difficult. For all these reasons, the trend in practical flying—as opposed to primary instruction—is toward increasing use of the power descent.

The power descent, as a flight condition, has been explained and illustrated in Chapter 2. A simple "clean" airplane such as a light plane, will invariably have its nose pointing slightly *up* in a power descent while its flight path is slightly *down*. A flapped airplane—whose power-off glide path with flaps extended is extremely steep, and whose nose in such a glide points very steeply down—will be in a level or slightly nose-down attitude during the power descent while its actual flight path is quite steeply down. In any case it is characteristic of this gait that the airplane is quite slowed up, so that it flies at high Angle of Attack. If the descent were made at high air speed and low Angle of Attack, with power on, it would be simply a dive and would be of little practical use, least of all as a landing approach.

Most airplanes are flown in a power approach at a speed quite a bit slower than their normal glide—perhaps halfway between the normal glide and the stall. But the speed depends on circumstances. A pilot will keep plenty of speed on in an unfamiliar ship, or in turbulent air, or on a ship which tends to turn nasty when slowed up. A pilot who is familiar with his ship and wants to make a very short landing in a small field will sometimes make a power approach at a speed only a few miles per hour faster than stalling speed. More than that: it was explained in Chapter 2 that some airplanes, especially multiengined ones, will have a slower stalling speed with power on than with power off. This difference is due to the propeller blast which keeps the stall from developing on those portions of the wings on which it would tend to develop first. If an extremely short landing is necessary, a pilot may therefore make a power-on approach at a speed at which with power off his airplane would actually be stalled! Then, when he finally has "dragged" it across the last fence, he needs merely to "chop" his throttles and the airplane drops to the ground right then and there, stalled and definitely "through flying." This is the method that results in the shortest possible landings. Needless to say it is reserved for pilots of experience, and pilots of experience have sense enough to use it only when the risks are justified.

The power approach has several advantages. It keeps the engine warm. It results in a slower loss of altitude than the power-off glide would. It brings the airplane to the ground along a shallower path, and hence makes the job of flaring out for the landing much easier; in fact, if enough power is used during the last stage of the approach a flare-out may become unnecessary, and the approach may simply be continued until the wheels touch the ground. At that moment the throttle is cut, the stick is pushed forward rather briskly, and thus a "wheel" landing is made. This is particularly useful in seaplane landings on "glassy" water and also in night landing on unlighted fields—in short, in any situation where one's exact height above the surface is hard to judge.

But the greatest advantage of the power approach is that the flight path can be so accurately controlled. Making gliding approaches

Glide and power descent with flaps—extremely exaggerated. In small light airplanes, the power-off glide is quite shallow, and in a power descent the airplane therefore is in a markedly nose-*up* attitude. *Left:* In a big heavily flapped airplane, the power-off glide is very steep. *Right:* Thus such an airplane is in a slightly nose-*down* attitude even in a power descent.

in an ordinary airplane you will be doing well—in actual practical operations, as opposed to flying-school "stages" and flight tests—if you don't overshoot or undershoot your intended landing spot by more than 200 feet; and every once in a while you may miss a whole airport! With power approaches, you can probably touch down within 30 feet of your intended spot, and you will never make those embarrassing extreme errors.

The control of the power descent is accomplished entirely by the throttle, and it is easy to understand. Exaggerating a little, one might say that the one extreme case of the power descent is the slow, mushing glide, that is, the slow-flight descent in which the power is cut all the way back. The other extreme of the power descent is slow, nose-high level flight—that is, a slow flight condition in which so much power is used that there is no descent. By varying your power between those two extremes, you can achieve any desired slope of descent between the steep slope of a slow mushing glide and the infinitely shallowed out descent of practically level flight.

There is nothing puzzling about it—at least not on paper. When you actually try to do one, you may get confused concerning the proper use of stick and throttle, and may wind up in a shallow dive or else in a power-off glide, or more likely, in an erratic, rollercoaster-like series of dives, zooms, mushing glide, and so forth. Hence, a note concerning the use of stick and throttle in the power approach.

The power approach is one extremely practical case in which it is definitely true that the stick is the airplane's speed control and the throttle is its up-and-down control. If you want to come down more steeply, *don't* nose the airplane down. Cut the throttle back, and then manipulate the stick so as to keep the air speed constant. As the throttle is cut, the nose must indeed come down a little if the speed is to remain constant. But in a perfectly behaved airplane, the nose will come down all by itself if you merely hold the stick steady! Since most of our airplanes are not perfectly behaved, you may have to take some action with the stick to keep the speed steady. But as a matter of fact, this action may well consist of increasing your back pressure on the stick rather than decreasing it! This is because with decreased power, most airplanes become disproportionately nose-heavy and actually want to pick up speed! At any rate, the thing to do is to free your working of the stick from any idea that the elevator is the airplane's up-and-down control. Work the stick so as to keep your speed steady. A steady air speed is the key to a successful power approach. Work the throttle to regulate your descent.

This is of course equally important in case you should find yourself rather low and want to descend less steeply. The "natural" tendency is to pull back on the stick and point the nose higher. This will then result in a loss of speed, and in increased mushing, and may actually fail to shallow out the flight path, except for a temporary ballooning. In addition, it will mess you up. Noting the very slow air speed, you will hastily advance your throttle, and the final result will be that rollercoaster-like series of ups and downs. No: if you want to shallow out your descent, simply open your throttle, and keep working the stick so as to hold the air speed constant. Here again this will probably require a slight *decrease* in your back pressure on the stick!

In talking this over with other pilots, you may sometimes be advised to use a different method: to put the airplane into a certain *attitude*, probably about level, and then work the throttle to regulate your descent—always keeping the ship in that same attitude. This method results, of course, in constantly varying air speeds and Angles of Attack. But note that in this method, too, you do not make the airplane go down by pushing the stick forward nor do you make it stay up by pulling the stick back. As a matter of fact, the handling of the stick works out just the opposite way! As you reduce your power to steepen your descent, you must come back on the stick if you want to hold the ship in its level attitude.

This constant-attitude method is second best. It is not quite as sure and smooth and foolproof as the constant-speed method outlined above. But it has the advantage that it is easier to teach to beginning students. A student often has no real conception of Angle of Attack and cannot yet sense it very well nor does he sense speed well. But he is invariably keenly conscious of his ship's attitude, and can handle the job of keeping the attitude constant, when he might fail at keeping speed and Angle of Attack constant. Also, it must be admitted that the constant-speed method outlined above depends rather heavily upon the air-speed indicator—since with power partly on, one's sensing of speed and Angle of Attack is not too keen. Now it may happen that a ship has no reliable air-speed indicator. Finally, many instructors think that the student's eye should be kept outside the cockpit as much as possible. Thus sometimes this constant-attitude method of making a power approach may be best.

PART V

GETTING DOWN

Once a pilot understands Angle of Attack and Lift, once he has right ideas, rather than wrong ideas, about his controls, once he can correctly execute the climb, the glide, the turn and straight and level flight—he should really have nothing further to learn: those things are all there is to flying, and he should be able to make his airplane do almost anything he wants it to do!

He should: but actually he doesn't. You know the difference between simply walking—and walking when it matters how you walk—for example, walking across a chasm on a narrow log, or walking across a stage in front of an audience. The same difference is found in regard to flying: the very great difference between simply being able to do the glide, the turn, and so forth, competently, and applying them to good advantage when it matters.

Now when it comes to the practical applications of flying, every student pilot knows certain examples: S turns across a road are a simple one, the pylon eight is a complicated one. But those are artifical, especially thought up applications; their only purpose is to afford student pilots a chance to apply their art. We shall not discuss them in this book. Those things are flight practice, and reading about them won't help you much; they must be practiced. Again, the military pilot and the air-line pilot and the test pilot could name practical applications of the glide, the turn, and so forth, which are rather difficult. But those are of no concern to the student pilot, or to the average commercial or private pilot. Hence we shall not discuss them either.

For the student pilot, there are really only two situations in which this business of the glide, the turn, and so forth, suddenly becomes real and earnest; and it suddenly really matters just how he flies. Those situations are the take-off and the landing. As for the take-off, it needs no special discussion in this book. What has been explained about torque and about the climb should suffice; practice will do the rest. But the landing is to most pilots the most difficult problem in the whole art of flying. And it is so damnably inescapable! Sooner or later, you've got to come down.

Thus, by a process of elimination, we arrive at the subject matter of the following chapters: how airplane and pilot behave between the time when the descent is begun and the time when the airplane finally comes to a stop.

Chapter 15

THE APPROACH

IF YOU gave a pilot an airplane and told him to practice up on his flying, this is almost certainly what he would do: He would go up to perhaps 1,000 feet, cut his power, glide in and try to land on a predetermined spot on the runway. And then he would go up and try the same thing again, and then again. In fact, for many a private flyer before the war, this was about all there was to flying—spot landings without help of power. Again, if you asked concerning a pilot whether he can handle a certain airplane, this is what most pilots would consider the test: Can he set it down, without help of power, where he wants to set it down? The same thing is very important in pilot's license tests: Anybody can fly an airplane after a fashion; what the inspector wants to know is whether the fellow can set one down without help of power on a prescribed spot.

The idea behind all this is, of course, that the engine may quit; and if the engine quits only an accurate spot landing will keep the pilot from cracking up.

An accurate spot landing is exceedingly hard to do. It requires so much practice that most pilots try to make every approach and landing without help of power, thus making every landing a practice forced landing. But just watch at any airport and see how many pilots need a last-minute blast of power after all to clear the wires that usually obstruct the approach to an airport; and how many need a last-minute slip to keep from floating clear across the field! And that is usually on a familiar field, under familiar wind conditions, and in the course of continual practice. The same pilot will do much worse at the end of a 3-hour cross-country flight, coming into a strange field. And the same pilot will probably do dismally indeed if his engine really does quit during a cross-country flight, when he perhaps does not even know the wind direction near the ground, let alone the wind velocity.

What makes it so hard? There is first of all the fact that the modern airplane has a very *shallow gliding angle* and thus must approach its field along so shallow slanting a line. It is as if you had to shoot at a target, with the target not facing you but set almost edgewise to you; the slightest error would make you miss not only the bull's eye but the whole target! In this respect, today's pilots have a much tougher task than the old-timer had in his less efficient, more steeply gliding craft; though, if equipped with flaps, a modern airplane can perform the steep glide that makes the whole approach problem much easier. Next there are the disturbances due to *wind* and especially due to updrafts and downdrafts. Even the best judged approach will miss if you hit a 300 feet per minute downdraft. Another difficulty is that the pilot has so little real *control* over the steepness of the glide path; you can't simply steer the airplane down if you are too high, and you can't stretch your glide very much if you suddenly discover that you are too low. But the main problem is the one of *perception:* Up there in the thin air with no reference points near by, it is simply very difficult to judge exactly where you are and where you are going.

WILL YOU NEED IT?

Fortunately, all this emphasis on the power-off approach is probably a bit overdone. Engines quit very rarely nowadays. Chances are that you will fly 10 years before yours will quit; and within those 10 years engines will be further improved; so that today's student pilot is extremely unlikely ever to have a forced landing. Moreover, engines almost never quit without warning. Finally, if an engine does quit without warning it is most likely to be on the take-off, because the strains on it are greatest then, and any faulty mechanical work will then have its first chance to show. And in that situation the pilot has no choice anyway but to go straight ahead and crash-land as best he can.

And the airplane itself will soon make the job of spot landing much easier. Flaps help a lot; but a few years hence, most airplanes will probably have even more efficient devices which will permit the pilot to steepen or flatten his glide at will, simply by pulling a lever or pressing a pedal.

Thus it would probably be better to put some of the emphasis and flying time on other things; other skills are more important for the pilot. For example, he ought to know how to make an extremely accurate approach and landing *with* power; for if he is ever forced down, it will probably be by weather or *gradual* engine trouble or poor navigation and impending darkness—his engine will be available to help. He ought to know how to navigate, how to judge weather, how to perceive drift. The accident record shows that pilots, as a group, are not good enough at making an ordinary turn.

But meanwhile it can't be denied that engine failure, though very unlikely, is very serious if it does happen, and that the accuracy of his power-off approach can thus suddenly become the most important thing in the pilot's life. Hence knowing how to do it will add to your peace of mind on cross-country even if you never use it. And also, there are those license tests to pass; finally, the same skill that goes into the judging of the power-off approach is also needed—in a lesser degree—in making an approach with power on. For all these reasons, the problem deserves thorough discussion.

Now a pilot's battle is half won if he really understands the glide and how the ship responds to the controls in the glide—that the way to make it go down more steeply is to hold the stick farther back, the way to make it go down less steeply is to hold the stick farther forward —just contrary to common sense. This will allow him to correct for errors of judgment as they become apparent to him. But there remains the other problem: how to judge.

JUDGMENT IS TEACHABLE

At one time, the whole art of flying was not considered teachable— it was thought to be a gift that could be cultivated only by much practice. We have long since discovered that this is not true. We break the art of flying down into small details that are teachable and learnable. We practice details. The details practiced are sometimes laborious, and sometimes they seem to lack practical value—for example, the coordination exercises, the chandelle, the various types of stalls. Yet we do practice them and combine them, and they finally make most of us pretty fair pilots. In the same way, the ability to judge an approach is even now not supposed to be teachable; but in fact, the art of judging an approach can be taught and can be learned. For it consists of tricks that can be taught and learned. Here, too, the details to be practiced may strike the student as tedious and not very realistic. Yet, if he would spend 10 minutes now and then observing how the eye really works during an approach, if he would watch the optical effects by which one can judge, if he would consciously practice the visual tricks described here, it would probably save him many expensive hours of just shooting approaches and landings.

All right then. Consider the thing first in its simplest form, the straight approach from 'way back. A pilot is flying toward the field, lined up with the runway, maintaining his altitude. Required: the time when he may close his throttle and just reach the intended spot in a comfortable glide. How does he judge?

Perspective helps in judging an approach. If you see the boundary lights of an airport as shown on the left, you know you cannot reach it in a glide. If you see them as shown on the right, you know that a glide, begun from your present position, might overshoot the field. If you see them as shown in the center, you know you are in a position for a straight-in glide. Correct perspective is different for different ships and different wind strengths.

The beginning student judges by absolute heights and absolute distances, and by memory. He remembers that, if he is to hit it right, he must glide past that certain tree at about twice the tree's height. He remembers that, in order to do so, he must cross a certain road at about the height of a fifth-floor window, with the grass in the fields looking thus-and-thus. And he knows that, in order to have the right altitude over the road, he needs about 500 feet on his altimeter when he crosses over a certain farmhouse.

This kind of judgment, by absolute heights and absolute distances, is all wrong. It will not work except on one's home field, in one's familiar ship, under familiar conditions of wind. Obviously it can't work on a strange field or if the altimeter isn't set to zero for the field or if there is a strong wind or no wind at all or if the pilot switches to a cleaner ship that has a shallower glide or, for that matter, if someone chops down that tree!

A better clue is the perspective in which the field appears: its foreshortened appearance as it lies before and underneath the pilot. This clue is used consciously by many pilots and unconsciously probably by all. In bringing a ship at night into a field that has only boundary lights, or only a flare path down the runway, it is sometimes the only clue, especially if the field is far away from towns or other lights and surrounded by darkness.

For the sake of simplicity, assume that the field is square. Then if it appears as almost square, you know that you are high over it and are thus overshooting. You know it even if you can see nothing else on the ground. If the square field appears radically foreshortened you know that it lies "in front" of you much more than "below" you: you are too low and probably can't reach it in a glide. If it looks "about right" you know you can probably glide into it.

This is a fairly reliable clue. It will work from any altitude, regardless of the absolute heights and distances involved; you get the same degree of foreshortening of a square as long as you view it from the same angle; whether you view it 5 miles away and 3,500 feet up, or 0.5 mile away and 350 feet up. Thus for one given airplane (and disregarding wind variations) there is one and only one perspective of the field that is "right"; it depends of course on the ship's gliding angle. A clean shallow-gliding ship wants its field sharply foreshortened, lying well ahead, and not too steeply below. A steep-gliding ship, on the other hand, wants its field much less distorted, lying well below and not too far ahead. A pilot soon remembers the particular perspective that goes with his ship's particular gliding angle.

The pilot in our example would first see the field extremely distorted; then, as he approached, losing no altitude, the distortion would become less and less, the apparent shape of the square field would gradually become more nearly square; and at the moment when the perspective looked "about right" to him, he would close his throttle. If his judgment has been correct, the glide will then bring him down into the field.

The field boundaries need not of course be actually square; the eye easily makes allowances for different shapes, uses paved runways, hangars, houses, in fact any familiar clear-cut object. And of course the pilot need not necessarily approach the field and watch the perspective steepen; maybe he glides about in the vicinity of the field, watching the perspective become more shallow; and then, when it is shallow enough to seem just right, he makes his turn toward the field. We are interested here not in the various approach patterns, but only in the clues by which approaches are judged.

The clue just discussed is good, but sometimes not quite good enough. If the field is located on a slope, as are many fields in the mountains; or if it is of quite irregular shape and featureless surface (snow, for instance) or if a seaplane approach is made to a large expanse of water, one's impression of the perspective slant becomes unsure.

But there are other clues.

Perhaps the most important of all is the position of the intended landing spot "under" the horizon. The pilot starts his glide when the spot is in the "right" position relative to the horizon. This will be described in detail; but it can't be made clear without first discussing the horizon, and how it appears to the flier.

THE HORIZON

The horizon is the line where earth and sky appear to meet. To an observer on the ground, the horizon always appears exactly at the level of his eye. That's why a level line is called "horizontal." As you climb up to altitude in an airplane and the ground gradually sinks below you, you would expect the horizon, too, to sink gradually below you. Certainly at extreme, astronomical altitudes you would see the earth as a globe, much as we see the moon, deep under you: the horizon then would have become a ring that would lie steeply below you.

That's what one would expect. What the flier actually finds as he goes up is that the horizon does not stay below: it goes up with him. At *any* eye level between the 5 feet or so of your own stature and the 50,000 feet of a stratosphere ship, the horizon is (practically) always as high as your eye. The line from your eye to the horizon is always horizontal. The edge of the world, instead of curving *down* away from you, seems to curl *up*. Instead of appearing like a globe under you, *convex*, the earth appears as a bowl under you, *concave*. You seem to be suspended above the middle of this bowl; and its rim, the horizon, seems to ring you all around!

Since there is no real difference between the groundman's view and the airman's view of sky and ground, this effect can really be noticed anywhere, anytime that you see the horizon. But on the ground, in the familiar environs of inhabited landscapes, the human eye is too sophisticated to notice it. The perspective of buildings and fields and trees, plus the fact that you *know* you are not in a bowl, destroys the illusion. But this bowl effect becomes quite strong, even on the ground, under unusual circumstances, especially to the more naïve eye of children or strangers. When Coronado first rode through the Great Plains where the air is clear and no trees hide the horizon, he reported to the King of Spain that in those parts one always had the impression of riding at the bottom of a valley, ringed by hills. Tourists from the country's interior, when they first view the ocean from the cliffs of Southern California, often remark that the water seems to be higher than the land!

One's reason says that the horizon should stay below; one's eye says that it comes up. The contradiction is explained simply enough: the earth is so huge, and the altitudes we can attain in airplanes are by comparison so tiny, that even a high-flying airplane is really hardly "up" at all! By scientific measurement it can be shown that the horizon does stay below. Navigators, for example, when measuring the "altitude" (angle of apparent elevation above the horizon) of a star, make an allowance for "height of eye," and in their very precise work it makes a difference of whether the navigator is at 5 feet, on the deck of a sailboat, or at 50 feet on a bridge of a steamer. Photographs taken from the stratosphere will sometimes show, on careful investigation, that the camera was looking very slightly "down" at the horizon. But at ordinary altitudes this "dip" of the horizon is so small that it is not apparent to the eye. Seen from 4,000 feet, for instance, the horizon has sunk below true eye level only by one degree!

THE HORIZON IS AT YOUR EYE LEVEL

For purposes of visual judgment in flying, then, the horizon is always as high as your eye. This comes in handy when flying near radio towers, mountaintops, or other airplanes: that which appears to you above your horizon is higher than you are. That which appears to you below the horizon is lower than you are. That which appears "on" the horizon (the horizon cutting through behind it) is at your altitude.

In the case of airplanes, this knowledge helps avoid collisions. In mountain flying, it cuts out unnecessary worry. The more distant mountains always seem to loom higher than the nearer ones, and one easily gets scared into desperate climbing, using up one's reserve of fuel, when as a matter of fact one could keep on cruising right across. Remember—if you can see the horizon above a mountaintop, then that mountaintop is lower than you are, however threatening it may look.

In the case of radio towers and similar obstructions, an understanding of these relationships is useful also on take-offs. If the obstruction sticks up into the sky before you, it is higher than you are at that moment, and you may run into it if you don't watch out. If the horizon appears above the top of the obstruction, you are higher than it is, and you will clear it. On a take-off from an obstructed field, the moment the distant horizon rises above the tops of the trees and other objects bordering the field, you know that you have cleared them at least, your eye has. Your landing gear, hanging down below you, might still tangle.

In this case, just as in the case of mountains, the inexperienced pilot is likely to focus his vision too anxiously on the obstruction alone, without regard to the horizon. If he does, the obstruction will always seem much more formidable than it is, and he may be tricked into climbing unnecessarily steeply, and possibly stalling.

That an object which appears above your horizon is higher than you are is true, incidentally, only of fairly near-by obstructions. A distant obstruction—say a mountain 30 miles away—may appear to stick into the sky, above the horizon, even though it actually does not reach up to your altitude.

THE EARTH IS A BOWL

And now, the bowl effect—which is of course merely a result of the horizon effect just discussed. Here we return to our discussion of how a pilot judges his landing approach. For it is the bowl effect that gives the experienced pilot the clues he needs.

The experienced pilot sees the ground much in the same manner in which the astronomer sees the heavens—in terms of "angular distances" rather than height, depth, and distance in the usual sense.

Here is what this means. You ask somebody, "I can't see that star you're talking about; where is it?" or "that bird" or "that airplane"; in short, you ask him the location of a point on the heavens. He will not be able to answer concisely. He will hem and haw and say "there" and point; he will make a wild stab in the case of an airplane

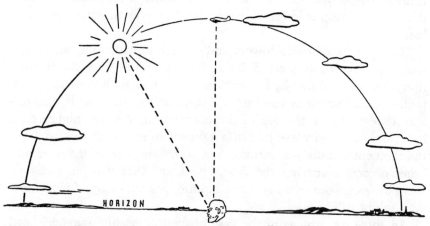

How we see the heavens. For many purposes, we consider the sky a dome on the inside of which the clouds, stars, birds, and airplanes are painted. Instead of describing the location of this sun and this airplane in terms of actual miles of distance and height—frequently impossible anyway—it is businesslike to say that the sun is about 60 degrees above the horizon and the airplane 90 degrees above the horizon; while the lowest appearing clouds which are actually simply the most distant clouds are only about 5 degrees above the horizon.

at guessing height in feet and distance in miles; or he will say, "It's high over *X*-ville." Not so the astronomer or the artillery man * or the mariner. He will say effortlessly and precisely, "The star is 15 degrees above the horizon" or, "The airplane is 45 degrees above the horizon and about 30 degrees to the right of those radio towers." He sees the heavens as a huge hemispherical bowl over him, on the inner surface of which the stars, clouds, birds, and airplanes are painted. It is obviously only a fiction, but it works; it allows him to measure and describe the apparent position of any point on the heavens with great accuracy.

* He uses mils instead of degrees. But it is the same thing.

The experienced pilot's eye sees the ground under him in a similar fashion: as if it were a huge hemispherical bowl under him, on the inside walls of which the towns and fields are painted. Consciously or unconsciously, the pilot is interested mostly in the *angle* at which things lie under him rather than in absolute distances and heights. Let an airport appear in the distance. The passenger might say, "The airport is about 3,000 feet below us but still many miles away." The pilot, in his own mind, will say the same thing differently, "The field now lies 5 degrees under the horizon." Later the passenger might

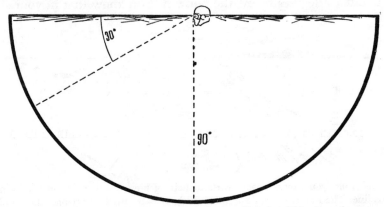

How we should see the ground. For many purposes, a pilot should look at the ground as if it were a bowl, the brim of which is the horizon. When a town appears below him in the distance ahead, it may be possible to say just how many miles ahead it is and how many feet below—and then again, it may not. But it will serve the pilot's thinking much better if he says: This town lies 30 degrees under the horizon.

say, "The height is the same, but the field is much nearer." The pilot will say, "The airport now lies 30 degrees under the horizon." He means that it lies more steeply under the airplane than it did before.

Both statements, the passenger's and the pilot's, express the same thing. But here is one difference: the passenger is only guessing. How does he know the distance, and the height? But the pilot is not guessing; although he doesn't know distance and height either, he does know that the field under him lies at that angle; he could even prove it, right from where he sits, by measuring the angle with some suitable instrument, such as a sextant. And here is another difference: The passenger's statement is useless; the pilot's statement is useful; it is angle, rather than actual height and distance, that matters. Here is why.

In a given ship, of given gliding angle, it is always the same point on the ground you can reach in a glide, regardless of your altitude; the same point, that is, in terms of angle-under-the-horizon. Say your ship's gliding angle is 1:5; this means you can in a glide always reach any point that lies 10 degrees under your horizon, or steeper. This statement (true only in still air) must be thoroughly understood.

Assume you are at 2,000 feet, somewhere over the country, looking at a field in which you intend to land. Can you reach it? You measure (mentally) its angular distance from the horizon, and you judge that it lies about 15 degrees below the horizon. You know that in your ship

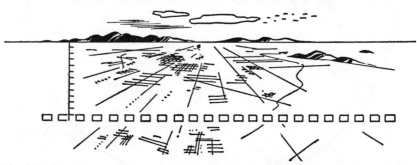

The Glide Line: you cannot reach, without help of power, any point on the far side of this line. The line is a certain number of degrees below the horizon—the number depends on the type of airplane.

you can reach any spot that lies 10 degrees below the horizon, or steeper. Hence you know that you can comfortably reach that field.

Now suppose that you spiral down several hundred feet, staying over the same spot of ground, merely losing altitude. Then look at the same field again. It now appears much closer to the horizon—only perhaps 7 degrees below it. It is now clearly out of your reach; for you know that you must glide at a 10-degree downward angle and can't glide any flatter. But now, at your lower altitude, it is still true that in your particular ship you can reach any field which now lies 10 degrees or more below your horizon.

THE GLIDE LINE

You know how the horizon appears from an airplane—a straight line all around, exactly at the level of your eyes. For purposes of learning or teaching how to make a precise approach, let us introduce the idea of a "glide line"—a line parallel to the horizon, but, say, 10 degrees (of angular distance) below the horizon. On the glide line are all those points of the ground which can just barely be reached in a glide by your particular ship. Any point that appears above the glide line is out of gliding range. Any point that appears below the glide line can be easily reached in a glide, will in fact be overshot unless the pilot "esses" or mushes or sideslips or otherwise steepens his glide.

And if you have understood what has been explained concerning angular vision, you will also understand this: How far the glide line lies below your horizon is entirely independent of your height; at any

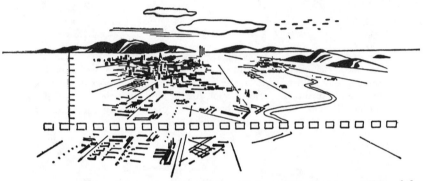

Imaginary glide line remains the same distance below the horizon regardless of the altitude from which you look at the ground, or, regardless of what part of the world you are in, whether the country is high or low, hot or cold. One of the first things to know about your airplane: how far below the horizon is its glide line.

height, the glide line is the same distance (*angular* distance, in terms of degrees) below your horizon. As your height changes in the glide, both the horizon and the glide line will be at different points in the terrain below you; but the horizon will always be at the same height as your eye; and the glide line will be the same number of degrees below the horizon; and the relation of horizon and glide line will not change.

How far the glide line is below the horizon does depend, of course, on the ship you are flying. If yours is a clean, shallow-gliding airplane, your glide line will be about 7 degrees below the horizon. If yours is a ship with a steep gliding angle, your glide line may be 15 or even 20 degrees below your horizon. But, wherever it may be, it will be that angular distance from the horizon regardless of your altitude.

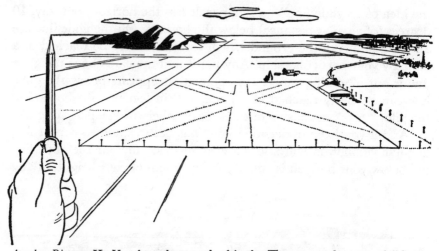

Angular Distance II: You have lost much altitude. The stream has moved "downward," out of your field of vision. The farm beyond the airport has moved "up" closer to the horizon. But the landing spot, just beyond the pole line, still lies at the same angular distance "under" the horizon, as measured by the pencil in your outstretched hand.

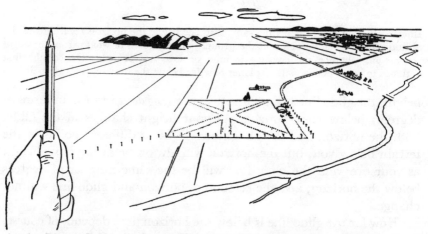

Angular Distance I. You are in a steady glide. The glide has been well judged and will reach the ground at the beginning of the runway, just clear of the pole line. As the glide proceeds, watch how the landing spot remains in the same position, relative to your horizon, regardless of altitude and distance.

All this seems no doubt very abstruse, and professional, and not worth the effort; but that is only because it is difficult to put into words and flat-paper pictures. Read this once at some high-up window where you can see the horizon; or read it in an airplane—and you will see that it is really quite simple and straightforward; it is, after all, only a formal description of what every experienced pilot actually does. And once you have understood it, you will have no trouble point-

Angular distance III. You are almost down, and still the landing spot has not moved, relative to your horizon. The farm, by contrast, is now almost on the horizon. *Your glide always will touch the ground at the one spot which does not move. It will always undershoot any spot which moves "upward," toward the horizon: it will always overshoot any spot which moves "downward," away from the horizon.*

ing it all out to someone else in flight—nor will he have trouble understanding it if it is shown to him in flight.

In our example, then, the pilot approaching in level flight from a distance would first see his field high up near the horizon, well above his glide line; as he kept flying, the field would gradually move downward, away from the horizon, until it reached and crossed the imaginary glide line; whereupon the pilot would know it was within gliding distance and would close his throttle. As he glided he would lose altitude. This would tend to make the field move upward toward the horizon and thus tend to put the field above and outside the glide line again. But he would also decrease his distance; and that would tend to make the field move away from the horizon and downward; if the glide had been started in the right place, the two tendencies would cancel each other, and the intended landing spot and the horizon would appear in the same position, relative to each other, all the way down. Try it!

Because this is so important, let's state it once more in different words. As an airplane is gliding toward a certain spot on the ground,

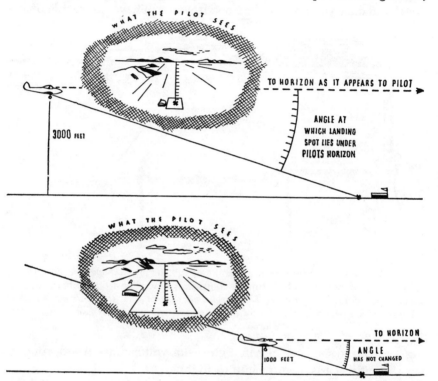

Why the spot does not move: as the pilot comes closer and loses altitude, the horizon comes down with him.

that spot, *as seen by the pilot*, remains always in the same relationship to the horizon, *as seen by the pilot*. The spot neither seems to move "upward" toward the horizon, nor does it move "downward" into a more steeply down position. It is stationary; throughout the approach it holds, to the pilot's eye, the same apparent location on the earth bowl.

And let's state it also as seen from the other end: as an airplane approaches a certain spot on the ground, it will appear to an observer on that spot as if it were not moving at all. It neither appears to move "up" toward the zenith, nor does it appear to sink "down" toward the horizon; it merely grows bigger: its apparent location in the celestial bowl remains constant.

In learning to fly, therefore, or in "checking out" on a new airplane, it is a good idea to make a conscious, systematic effort to find out: how steeply below the horizon lies the spot which this particular airplane will always reach in a glide?

THE APPROACH AROUND THE CORNER

In practical flying, even this is not good enough. The pilot might be on an unfamiliar ship, whose gliding angle he doesn't know; winds of unknown and variable strength might make the remembered gliding angle illusory. In a real forced landing, this is the main difficulty. Sometimes the pilot does not even know the direction of the wind near the ground, let alone its velocity. And finally, visual judgment and visual memory are deceptive.

In music, it is the exceptional individual who has the gift of "absolute pitch," that is, the ability to sing an intended note correctly or to tune a violin correctly without guidance by a few notes from some fixed-tune musical instrument. In flying, visual judgment is often similarly in need of being "retuned" before the performance. After a long flight at high altitude, one tends to misread one's optical clues, judge oneself lower than one is, and overshoot.

To overcome all those sources of error is the purpose of the 90-degree approach. The pilot glides along the down-wind side of the intended field, at right angles to the wind, neither substantially increasing nor decreasing his distance from the intended landing spot, but of course all the time losing altitude; and he continues to glide in that manner, until he judges the right moment has come to turn in and make his final glide down to the spot.

The functioning of the pilot's visual judgment during such an approach will now be described in detail. It is assumed that the pilot approaches the field in the counterclockwise manner usual for airport traffic. The so-called "base leg" of the approach is made with the airport lying on the pilot's left quite steeply below him at first; and the wind coming from the pilot's left. While gliding along this base leg, the airplane is of course losing altitude constantly but is neither increasing nor decreasing its horizontal distance from the intended landing spot—at least not substantially. Hence the spot moves gradually higher and higher up toward the pilot's horizon, and when it has moved high enough—when it lies at a flat enough angle—the pilot turns left and heads directly for it. The key to the whole maneuver is the timing of this final left turn. If it is made too early, the landing will be overshot. If it is made too late, the landing will be undershot.

All the judging is done during this base leg of the glide; for, once the pilot has made that final turn toward the spot, he is committed and has little further control of his exact landing spot. This judging is done in four separate and distinct optical operations, two that *collect* information, and two that *apply* the information thus collected.

COLLECTING THE INFORMATION

In operation number one, the pilot finds out his ship's gliding angle. This is done simply by watching the ship's motion and determining the point on the ground at which it would arrive if the glide were continued in that cross-wind direction. Since at this time the ship is gliding at right angles to the wind, the wind neither steepens nor flattens gliding angle; what the pilot observes in this fashion is therefore essentially the ship's gliding angle in still air. And this is one of the main advantages of this type of approach: it allows the pilot to refresh his memory, as it were, of his ship's gliding angle—as the singer's memory of pitch is refreshed by a few notes sounded on the piano.

Operation number two is to look over to the left, toward the intended landing spot, and mentally project the gliding angle just established into that direction. It then becomes evident whether the glide, if turned toward the spot at that moment, would overshoot the spot or just reach the spot or undershoot the spot.

This is not a difficult pair of operations, yet there seems to be an efficient way of doing it and an inefficient way. The inefficient way is judgment by absolute heights and absolute distances and familiar objects, while the efficient way uses the horizon, angular measurements, as discussed earlier.

Let us try to transcribe those quick and half-conscious mental processes into speech. The less skillful pilot thinks, " . . . seems that my present glide would get me down about two miles from here, down near that farmhouse with the red barn. Therefore, if I turned left now, I would also glide about two miles; that would get me—now let's see—" (and here he turns his head over toward his intended landing field) "that would get me 'way beyond my intended field. I had better wait." He waits, and after a while repeats the two operations, until finally it seems to him that his glide would take him to the intended spot. At that time he makes his turn, and if he has judged right, he just makes it.

This is not, however, the efficient way. In the first place, one's judgment of distance is not reliable enough; it is influenced by such things as light, haze, the kind of objects one is looking at. In the second place, the pilot who judges in this way has to repeat this whole procedure continually—for, as he is losing altitude in his cross-wind glide, the distances involved are of course constantly changing.

Here, by contrast, is how the more efficient pilot would think, " . . . seems I am moving right toward the point there on the ground where that farmhouse—" and here he interrupts himself, "but who cares about farmhouses; I am moving right toward a point which lies— let me see—how many degrees under my horizon?" And here he measures mentally the angular distance between that point and the horizon. How he measures it is a bit hard to describe. Prof. David Webster uses the very descriptive phrase that he "drops his eye," that is, he looks at the horizon and then looks downward at his point (the farmhouse), and fixes in his memory the amount of eye motion necessary to shift his glance from horizon to point. This is probably what actually does happen. If an instructor wanted to dramatize the process and make it quite visible, he might take a pencil aloft with him and use it, with outstretched arm, to sight over—in the manner in which an artist sights over his pencil to make approximate angular measurements while sketching.

This done, the pilot turns his face left, toward the landing field, looks at the horizon, "drops" his eye through the remembered angle, and thus finds the point at which his glide will bring him down. Or our dramatizing instructor would hold out his pencil to the left, sight with its help, and thus measure off the angle at which the glide would bring him down.

Once the pilot has determined the point, in terms of angular distance below the horizon, which he can reach in a glide, he can now simply wait and watch the landscape below distort itself and assume new perspectives. Gradually his intended landing spot moves up

Approach around a corner II. Note: *This* is Part 2; Part 1 is on the *right*. First: Working "downward" from the horizon, find point *C*—the point at which you would touch down if you turned toward it *now*, and if there were no wind. You get this point by "dropping your eyes" through the same angular distance through which you dropped them in Part 1, between the horizon and point *O* there. *Next*, starting from point *O*, working toward yourself, apply drift distance *OX*, of Part 1, but apply it properly foreshortened by perspective. The result is point *X*: at this point you would land if you made your turn into the wind at this moment. Obviously, you are still a little high and would land on last third of runway instead of first third as is good practice. Wait about 15 seconds, *then* turn.

Approach around a corner I. Note: *This* is Part 1. Part 2 is on the *left*. You are in a glide, looking over the nose. The airport is on your left. The wind is from your left. *Required:* The time to turn left so as to land on desired spot. *First,* watch your glide; find the spot which moves neither up nor down—the spot on which you would land if this glide were continued all the way to the ground, and if there were no cross wind. This spot is marked O. *Next,* watch your sideways drift; find the point to which you would drift if the glide continued. This spot is marked X. Remember the *angular* distance at which O lies under the horizon. Remember the *actual* distance, in feet or in quarter miles, between O and X.

Then look to your left.

nearer and nearer to the horizon; and at the moment when it has reached proper nearness to the horizon (when it lies at the angle at which his glide will bring him down), at that moment he turns in toward it, and just makes it.

If there is no wind, then that is all there is to it. But if there is appreciable wind, the pilot must also gauge its velocity and make allowance for its effect. This is the purpose of operations three and four. During the first leg of a right-angle approach, the wind is across the flight path, and the resulting sideways drift gives the pilot a clear impression of its velocity; and he can make some allowance for it by guess. It consists of making the final turn to the field more or less ahead of the time that would be right in still air, that is, the pilot turns while still in position which in still air would cause him to overshoot; and he depends on the wind to steepen his final glide.

This gauging of wind effect can be taught and can be learned by a little systematic practice. In this instance, the pilot's vision does not function like the astronomer's, in terms of angles and degrees of arc; in this instance, it is interested more in actual miles and feet of distance.

Operation three, then, goes as follows: As he looks forward he can see, not only the angle at which the wind is drifting him sideways, but also the actual distance that he is going to be drifted down-wind; for instance, he is *headed* exactly for that farmhouse with the red barn; but he can see that if he continued the glide all the way down, he would not actually get to it but would touch down a quarter of a mile down-wind of it. He knows now that this quarter mile is the amount of drift which he will suffer between now and the time when he arrives at the ground.

And now, operation four: This quarter mile the pilot's eye takes and carries over to the left; if he is going to be set a quarter mile down-wind between now and the time he gets to the ground, he must now gauge his approach as if he wanted to land a quarter mile up-wind of where he actually does want to land; for then the quarter mile of drift will just bring him down on his intended spot. Thus he measures off with his eye a quarter mile up-wind from his intended spot, to find the new spot at which temporarily to aim his approach. In measuring this off, he must of course make allowance for the perspective foreshortening in which that part of the ground appears to him. When he looks forward and estimates the distance he is going to drift, he sees the quarter mile undistorted, in its real size; when he transfers it over to his left and measures it off up-wind from his intended spot, it looks shrunken and small. But his eye knows from ground experience how to make that sort of allowance properly. It is not difficult and after some practice becomes automatic and results in accurate approaches.

This estimate of drift distance is good, of course, only for the moment at which it is made. It must be remade continuously until the turn-in is finally made.

GOOD FORM OFTEN SEEMS AWKWARD

What has just been described makes rather hard reading. It is the sort of thing which the average man does not want to visualize through his own mental effort, but would rather be "shown" by an instructor. The last few pages have been aimed, perhaps, largely at instructors. Once he has grasped them, the instructor will find it rather easy to point these things out to the student in actual flight. The glide should be begun several thousand feet above the ground so that there is plenty of time for explanation and observation. And it may be a good idea to point these things out to a student first of all while standing in the airport control tower or some other high place which is not noisy and bumpy.

The student pilot, however, will find that these things are not really so hard to grasp. A quarter hour of concentrated reading, given these last few pages, preferably in some high-up place where the horizon is clearly visible, will make them clear. And once grasped, these tricks of visual judgment are really quite simple.

Even then, though, they require practice. At first, the more formal and systematic method of judging which is described here will seem rather clumsy, unduly hard work, and less efficient than the freehand style by guess and by gosh. But this initial awkwardness is found also in other phases of flying, and in all other sports or skills, whenever an attempt is made to achieve really good form. The skilled breathing technique of the singer, for example, seems at first quite awkward and *is* at first quite awkward. But in the long run the singer will be better off because of it.

THE FINAL GLIDE

Now for the final glide. You have made your judgment and **have** turned toward your spot. But even at best, there is a probability **of** error having crept into your judgment, and there are also updrafts **and** downdrafts to spoil your accuracy, and variations of wind with altitude. During the final glide you are bound to do a good deal of worrying, especially if it is a flight test or an actual forced landing. Are you going

WILL I MAKE IT?

Will I clear the obstruction? Note the tuft of grass on runway, just "above" the pole. It is the reference point by which you can judge.

to make it? Thus you must keep making visual judgments. And you must make them fast. For overshooting is much easier to correct without violent maneuvering, if it is recognized early; undershooting can be corrected, without use of throttle, only if it is caught quite early.

Are you going to make it, or aren't you? How can you judge?

One clue now is the apparent motion of objects on the ground. As you look forward-downward at your intended spot, two optical phenomena are happening, both at the same time. In the first place, all objects are growing in apparent size, because they are approaching closer to your eye; the runway grows, the hangars, the trees, the field beyond the airport, everything comes at you and grows. But at the same time, most of the objects, as they "grow," also show some motion; they move to different places in your field of vision.

Some move downward. As an example, take the airport fence. In the first stages of the final glide, it is practically in front of you; then it

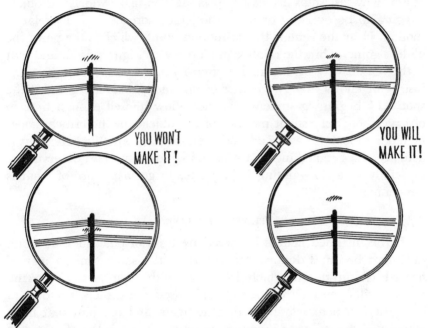

Left: If the obstruction seems to grow "up" into the field so that its top appears "above" the original reference point, your present glide is carrying you into the pole line. Take action. *Right:* If the obstruction seems to "retract" out of the field, pulling "down" from the original reference point, your present flight path will carry you safely across. Sit tight.

moves downward in your field of vision until it finally flashes by under you. Some objects move upward in your field of vision. As an example, take some field a little way beyond the airport. At first it lies practically in front of you, and your vision encompasses it easily along with the airport itself. Toward the end of the approach, when you look at your intended landing spot, you would have to raise your eyes quite noticeably to see that field.

Thus the objects move. The clue that tells you whether you are going to overshoot or undershoot is this: All objects that move downward, however slightly, are going to be overshot; all objects that move upward, toward the horizon, however slightly, are going to be undershot. And the objects that remain stationary in your field of vision and just grow in apparent size—those are the objects you will hit.

The faster the airplane, the more lively is this clue. This is one reason why fast ships are easier to land. The more widely an object is going to be overshot or undershot, the more lively is its relative motion. Near the center, the no-motion point, which is the point the glide is going to hit, the relative motion is very slight indeed; and that is the reason why beginners are often flying in doubt long after the instructor knows exactly whether or not the glide is going to hit the spot. The beginning student does not allow himself enough time to observe those very slight motions of the objects in the critical zone; he tends to look for a second and then look at something else; but it actually takes something like 4 or 5 seconds of patient observation, especially in a slow-gliding light plane, to get a picture of what is happening.

THAT POLE LINE

If the intended field is bordered by trees or pole lines, accurate judgment becomes doubly important; at the same time, a mental hazard is introduced which interferes with judgment. Often the student will open the throttle to help himself across the wires, only to have the instructor close the throttle again and tell him that he is going to clear easily. As judgment improves, there is less of that, but even so, obstructions on the approach are unpleasant.

Here is how you can judge with great precision. Approaching the field, you see the tops of the poles projecting into the landing field. That is, you see the top of a pole against a background of airport surface with tufts of grass, or bare spots, and so forth, providing convenient reference points. Now, if during the further approach the tops of the poles are apparently *growing* deeper into the field, however slightly, you are not going to clear the line. If the top of the poles seem to remain fixed, relative to their background, you would clear if you hadn't a landing gear; but your landing gear will tangle. But if the tops of the poles are apparently *pulling out* of the field, however slightly, you will make it.

Chapter 16

THE LANDING

IN TRYING to understand how airplane and pilot behave during a landing, we had best begin at the end. The whole landing maneuver is shaped toward one final instant, one final requirement: the instant of contact with the ground, and the requirement that the contact be of exactly the right sort. In other maneuvers, the pilot can continually correct his mistakes as they become apparent to him. In a steep turn, for instance, the pilot's whole action is nothing but a series of small corrections. In the landing, the error becomes apparent often only upon contact with the ground, at the instant when it is too late for correction. Accordingly, the student should first of all thoroughly understand the exact manner in which the airplane makes contact with the ground.

WHY THREE POINTS?

The conventional way of landing is still the three-point landing. In such a landing the airplane touches the ground in a nose-high attitude, and touches it with all three points of support, main wheels and tail wheel, simultaneously. Why this attitude? And why the three-point requirement?

It is sometimes said that the landing maneuver is a stall, brought on a foot or so above the ground. This may be so, but it is not necessarily always so. Quite a few airplanes touching the ground in a normal three-point landing are not stalled. At any rate, the stalling is not the essence of the maneuver; the essential part is slowness. The designer is anxious to make the landing of his airplane take place at about the slowest forward speed at which the ship, with its given wing and its given weight, is aerodynamically capable of flying.

He wants that slow landing for several reasons. In the first place, landings at higher speed would impose greater stresses on the ship's structure, forcing him to use a heavier structure; this would cut down the ship's useful load. In the second place, the designer usually believes that the slower a landing is, the easier it is for the pilot to execute. This belief is largely erroneous, but it is certainly very generally accepted. Finally, the designer has in mind the possibility of an emergency landing on a small field; and the landing run of an airplane increases "as the square of the speed," that is, if you double the landing speed you make the landing run four times as long. He has in mind also the possibility of a landing on rough ground, of a crash landing, or of a landing accident, in short, the possibility of the ship's hitting something; in that event, the destructiveness of the impact, and the danger to the occupants, also increases "as the square of the speed."

A three-point landing (shown here is an extremely "floating" one) is made by flying the airplane level just above the ground, more and more slowly, at higher and higher Angle of Attack, until stalling Angle of Attack is reached. It then sinks . . .

In all those respects, lowering a ship's landing speed from 56 m.p.h. to 40 m.p.h. amounts therefore to cutting it in half! It halves the kinetic energy involved, the distance required to come to a stop, the risk of injury, the viciousness of any crash.

For all these reasons, then, the designer wants the ship to make contact with the ground while in very slow flight. In very slow flight, an airplane must be held at high Angle of Attack; hence that characteristic nose-high attitude, and hence the requirement that the stick be all the way back, or almost so.

As for the three-point requirement, it is quickly explained: the designer simply proportions the length of the landing gear legs and the height of the tail wheel support so that, when the ship flies close to the ground in nose-high slow flight, all three points will touch the ground simultaneously.

The exact length of the landing gear legs depends therefore on just how much the designer expects the pilot to slow the ship up before setting it down. In most airplanes, the designer goes pretty near the limit in this respect; he expects the pilot to slow the ship up to actual stalling speed before letting it touch the ground, and he proportions his landing gear accordingly. But there are exceptions; some perfectly normal airplanes are set on landing gear with rather short legs and a comparatively high tail wheel, and are meant to be landed unstalled, that is, not slowed up to the extreme. If you land such an airplane three-point, it lands several m.p.h. faster than the slowest flight which it is capable of maintaining. If on the other hand you land it as slowly as possible, in a stall, with the stick all the way back, you don't land three-point at all but touch the ground with your tail wheel first, while

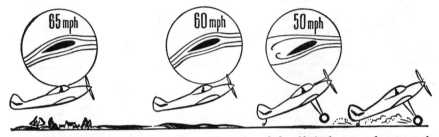

. . . to the ground. Landing gear is so proportioned that if airplane touches ground in this extreme slow-flight condition, all three points will hit at once.

your front wheels are still 9 inches in the air. Again there are airplanes, built for military liaison work, which can be flown at unusually high Angles of Attack, because of slots or other antistall devices on their wings; in order that such airplanes can take advantage of their slow-flying ability, they need of course unusually long landing-gear legs and must sit on the ground in an unusually nose-high attitude.

To sum this up, then: It is not really true that, when we land, we pull the airplane's nose higher and higher in order that we may hit on all three points at once. The true story is the other way 'round. The landing gear's three points are where they are in order that they may hit the ground simultaneously if the airplane approaches the ground nose-high in extremely slow flight!

For the beginner, it is no simple task to fly the airplane onto the ground in very slow flight or actually to let it sink onto the ground in a stall. It requires a visual judgment that comes only with practice. And it feels wrong at first, for it also requires that the stick be held well back and brought back farther and farther, until, in a full-stall landing, it is hard against the pilot's stomach. So close to the ground, that goes very much against the student's nervous grain, and rightly so. Almost every student, reluctant to get the stick as far back as is necessary, fails to get the ship as well slowed up as the designer intended it to be, hits with his two main wheels first, and bounces.

BOUNCES

Three separate effects act on an airplane in such a wheel-first landing. Two are unfavorable, making for a bounce; one is favorable, tending to inhibit a bounce.

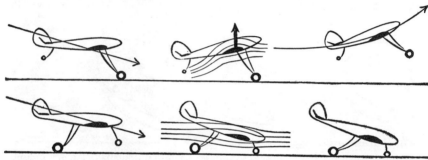

Top: On *conventional* landing gear, airplane tends to "bounce." Ground contact, unless made either "three-point" or extremely gently, tends to nose ship *up*. Angle of Attack thus increases, a temporary surge of new lift carries the airplane away from the ground. Lift surge then dies, and heavy drop follows. *Bottom:* On *tricycle* landing gear, airplane hugs ground. Whatever the speed and attitude of the ship, ground contact noses it *down*. Angle of Attack decreases, may even become negative. Lift is reduced, or may even become negative and press the airplane down.

The first of these is the *ground impact* effect: As the ship settles to the ground and hits front wheels first, the nose receives an upward shove at a moment when the tail has not yet found the ground and is still sinking. The result is that the airplane noses up sharply: the Angle of Attack of the wings increases: their lift increases. Exactly as if the pilot had suddenly yanked back on the stick, the ship balloons upward.

This is called a bounce, for it looks exactly as if the elasticity of the landing gear had caused the ship to rebound back into the air, in the same manner in which a ball bounces. Actually, it is not a bounce but an involuntary take-off: the ground impact merely disturbs the ship's attitude a bit; the force that actually lifts the airplane away from the ground is not the ground impact, but the action of the air upon the wings as they suddenly assume a much higher Angle of Attack.

The whole effect is not so vicious, however, as many pilots think it is. Its viciousness depends on a number of factors. One is the stiffness of the landing gear. If the gear is bouncy, ground contact will naturally shove the nose up hard; the wings' Angle of Attack will increase suddenly; there will be a sharp lift and a marked bounce. But most landing gears now incorporate, in addition to the actual spring or other elastic element, some type of shock absorber which enables the landing gear to "take" a sharp impact readily while paying it out again only slowly; the landing gear leg "gives" almost instantaneously upon impact, but its subsequent elastic rebound is retarded by various frictional devices. With such gear, ground contact with the main wheels first does not give so hard an upward shove to the nose, and all the subsequent effects are much gentler.

Another important factor is the sinking speed of the airplane just at the moment of ground contact. A tense or confused beginner will sometimes freeze toward the last of the landing, when perhaps 5 feet off the ground. He stops his gradual pull-back on his stick at this point, fails to do anything further to check the ship's descent, but simply sits there with the stick held halfway back waiting for ground contact. Naturally, the airplane will begin to sink; and naturally, if this sink goes unchecked all the way down to the ground, it will become quite pronounced, and ground impact will then produce a rather sharp nosing up of the ship.

The more experienced pilot, too, frequently misjudges his height slightly or misjudges his rate of descent or, in a strange ship, misjudges the height of his landing gear. Thus he, too, frequently makes an inadvertent wheel-first ground contact. But, since he does not drop in from several feet, but is up to the moment of actual ground contact engaged in coming back on the stick and slowing his ship up and thus keeps his sinking speed under control, the resulting "bounce" is shallow and harmless and may even fail to materialize altogether.

SLOW THINKING

The second of the three effects that mark a wheel-first landing is the student's delayed reaction on the stick. This is really what Hollywood calls a "double take": the delayed and inappropriate reaction of a man who doesn't "catch on" to a sudden change of the situation and keeps bumbling on as if nothing had happened. In the movies, this is usually a comic effect. In flying, it is the main cause of those steep, leap froglike jumps that are so characteristic of student landings.

It happens about like this: The student has been told dozens of times that the stick must be all the way back in his stomach at the moment when ground contact occurs. He has failed several times in a row to get the stick all the way back. This time, he is determined not to be caught napping. But again, before he has had time to follow out the good intention, he feels the rumbling of the wheels. At that moment, he yanks the stick all the way back in what he thinks is the nick of time: "just caught it," he thinks.

It comes too late. Since ground contact has already occurred, it not only does no good but does positive harm; it merely serves to nose the ship up with the flippers at the very moment when the ground impact is also exerting its nosing-up effect; the ship noses up twice as sharply, and the ensuing bounce is twice as steep as it would have been had he done nothing at all!

It is probably this effect, rather than the faulty ground contact itself, that causes the typical student bounce. To get rid of it, the student should make up his mind beforehand, not merely that he will get the stick back as far as it will go, but also that he will arrest his hand the moment he feels the ground under his wheels. At present, it is accepted practice to hold the stick all the way back during the landing run. If it were not, it might sometimes be a good idea if a student let go of the stick altogether at the moment of ground contact.

The third of the three effects that mark a wheel-first landing is the effect of *ground friction*. This is the favorable one; it reduces the tendency to bounce. The friction of the ground, tending to hold the wheel back, acts through the leverage of the landing gear legs, and thus tends to pull the ship's nose down. Depending on the nature of the landing surface and on the speed, this friction effect can be quite faint, or it can be extremely strong. In at least one case it is suspected to have been strong enough to cause a fatal accident. At an air show a pilot touched his wheels to the ground after diving the ship to excessively high speed. The drag of the landing gear seems to have pulled his nose down so forcefully that, presumably with full back stick, he was unable to break free of the ground and finally nosed over.

A rather startling demonstration of the same effect can be put on with an ordinary trainer on a grassy field. You can glide it into the ground without any attempt to set it down carefully, three-point, two-point, or by any other system: simply let it hit, and at the moment of ground contact, take your hand off the stick altogether. The ship will not bounce, nor will it nose over. It will hug the ground nicely while it rolls tail high across the field at high speed. As the speed slackens, it will drop its tail and from then on will decelerate rapidly in the usual manner. On a grassy field, this will work whether you hit slow or fast—up to cruising speed—whether you hit power on or power off, steeply or shallow. All that is necessary when trying the maneuver is to break the glide once so that the ship glides into the ground at about half the angle of its normal glide; then, without any further manipulation of the stick, let it hit. The breaking of the glide angle is necessary not because the ship would jump or bounce upon steeper impact, but merely because the impact itself becomes rather severe on the ship's structure when it takes place at too steep an angle.

On a paved runway the same sort of ground contact will probably result in a bounce. The pavement does not develop enough friction to hold the nose down firmly enough. But even on such a runway, the bounces are less severe than most pilots would expect.

Altogether, then, it would seem the effect of ground friction (which tends to keep the ship from bouncing) counteracts most of the effect of ground impact (which tends to make the ship bounce); and it seems that the typical student bounce is caused mainly by the delayed reaction of the student, his jerking the stick all the way back after ground contact has been made. Moral: Don't.

What should the pilot do if he does bounce? Probably all pilots actually do the same thing; but the descriptions they give of that thing are not the same. And here, or elsewhere in flying, a slightly different wording of instruction, a slightly different concept in the learner's mind, can produce amazing differences of behavior.

One pilot might simply tell you, "Keep flying it. As you realize that you have bounced off the ground, push the stick forward to keep the bounce flat and stay at least close to the ground if you can't actually stay on it. And then, as the ship again approaches the ground, start coming back again on the stick and try for a three-point contact." But, if a student tries this, his control action is almost certainly going to be out of phase with the action of the ship. His forward pressure is likely to take effect just at the moment when the ship reaches the top of the bounce anyway and is already being nosed down by its own inherent stability. The result will probably be that the second ground contact is even more "on the wheels" than the first one was, and a second steeper bounce will result.

Another pilot might tell you, "When you bounce, don't do anything. Hold your hand steady and watch; and when you get to the highest point of the bounce and begin to approach the ground again, resume your pull-back on the stick, and this time get it all the way back." This is a pretty good recipe, especially for a lightly wing-loaded airplane and for a normal, not too severe bounce.

Still another pilot might tell you this, and it is probably the best recipe and works on airplanes of all sizes and wing loadings, "If you bounce, concentrate your attention on the *attitude* of the airplane. Do with your stick whatever is necessary to put the airplane into a three-point attitude, that is, into the attitude in which it sits on the ground; and, once you've put it into this attitude, do with the stick whatever is necessary to hold it there. And, if the bounce was at all severe, be ready with your throttle hand to feed a blast of power, to keep from dropping in too hard."

THE FLOATING LANDING

Now for a close look at the maneuver which brings the ship down to the ground on its three points. This can be done in two different ways —the "floating" landing and the "stall-down" landing. To prevent confusion, let's state once more that those are two different ways of getting the airplane down for a three-point contact with the ground. The stall-down landing is more difficult, more precise, and more proper—it is the way the Army prefers in primary training. We shall discuss the floating landing first because it is easier to do and much easier to understand.

The pilot approaches in a normal glide and levels out only when quite near the ground; so that he finds himself shooting along level, a foot or two off the ground, still with plenty of excess speed. The process of landing then consists simply of holding the ship off the ground as long as possible. Because the engine is idling and the flight path is level, speed will naturally slacken. As the speed slackens the lift decreases; the ship will sink through to the ground (making a bad, wheels-first contact) unless the pilot comes back on the stick, pulling the nose up and forcing the airplane to higher Angle of Attack. The higher Angle of Attack then restores the lift, so that the ship keeps floating. But presently the speed slackens again, and again it will sink unless the Angle of Attack is again increased by pulling the stick farther back. The higher the Angle of Attack, the more rapidly does the speed slacken; the more rapidly the speed slackens, the faster must the pilot increase the Angle of Attack to keep from sinking to the ground; so that this "hold-off" requires a continual backward motion of the stick, which is at first almost imperceptibly slow, then becomes faster, and finally in some ships ends up as almost a yank that brings the stick back into the pilot's stomach. When the stick is all the way back, there is nothing else the pilot *can* do to hold the ship off the ground; it will sink to the ground. And, as described above, its landing gear legs are so proportioned that, as it sinks down in this flight condition, it is bound to hit on all three points at once.

DEPTH PERCEPTION CAN BE LEARNED

But how is the pilot to know just at which height to level off, and just how high above the ground he is?

This is where the awesome subject of depth perception comes up. According to the fit-to-fly books, this judgment is based upon depth perception, the mysterious ability which your doctor tested when you had those reins in your hand and had to move those two pencil-like objects in the little puppet-show stage. Depth perception is the result of your having two eyes set in your head some distance apart. In looking at a thing, you become the more cross-eyed, as it were, the closer the thing comes to you. Also, each eye sees a slightly different picture from its slightly different point of observation. All this translates itself in your brain into a perception of distance—or depth.

But this sort of depth perception—let's call it "pure" depth perception—has nothing to do with landing or indeed with any phase of flying—except formation flying. That it is important, or even indispensable for a pilot, is merely one of those old superstitions which writers on aviation copy out of each other's books, just like the proposition that the rudder is the airplane's directional control or that pulling the stick back makes it go up. Except for formation flying, the only occasion when depth perception of the pure sort can be used in handling an airplane is on the ground, in taxiing close to hangars, other airplanes, etc., when the pilot must judge whether his wing tips will clear. And an examination of wing tips shows that most pilots are not equal to the task; the air lines therefore have long painted white strips on the airport ramps to keep their pilots from bunging up their airplanes!

In landing, the pilot's vision is not directed downward and is not even attempting to perceive the depth yet remaining between him and the ground. It is directed well forward at a region hundreds of feet ahead of the airplane—well beyond the reach of pure depth perception.

Depth perception of the direct kind is reliable only through quite short distance; beyond 100 feet or so it fades out entirely, and you can no longer judge depth in this fashion. If you do perceive depth beyond those distances, you do so by different means—all indirect: the haziness of things; or, if they are things of known size such as trees, their apparent size; or, if they are seen from a moving point of observation (perhaps from a train window) the speed and manner in which they appear to move past one another. You notice that all these ways of judging require experience. The Easterner misjudges depth when he first gets into the clear air of the West. Or, if the distant trees that you thought were ordinary pine trees are actually giant redwoods, they are farther away than you thought.

Thus, since all the clues of which a pilot can judge depth are of the kind that require experience, even a person with perfect vision must learn them—learn them formally or informally.

On the other hand, even a one-eyed man (who can have no "direct" depth perception) can easily learn how to judge a landing. You can prove this to yourself by landing with one eye closed. And it is proved also by the whole career of one of the greatest pilots of all times—Wiley Post. Like so many things in the supposedly elusive art of flying, the judgment of the height in landing can be broken down into teachable, learnable detail. It becomes quite easy if you only know what to look for. In fact it then becomes an almost mechanical trick.

HOW TO JUDGE A LANDING

As you approach the ground you must keep your vision relaxed and look all around; you must take in the whole scenery, the perspective of the hangars on the side of the field and the other airplanes on the field, the parked automobiles, the trees, the telegraph poles all

Depth perception? In landing, height above the ground is judged not by "depth perception" but by perspective in which things appear. Each of the objects pictured can appear, in the particular perspective shown, only from one particular height—a different height for each of the objects shown. One single glance tells you in each case how high you are above the ground or water.

around, the grass, the horizon; for it is from the perspective and apparent motion of such things that you will get a vivid perception of your height; and a staring eye will not see what matters. When you get tense, you will almost certainly stare; approaching the ground, most students do get tense: that is largely why the landing is so difficult for most beginners

Your first problem is to know when (in the glide) the time has come for you to "flare out," that is, stop the descent and make the airplane float level. Of course the breaking of the glide, the bending of the flight path from a downsloping to a level direction, requires a certain amount of time and space; it can't be done instantaneously in a sharp corner. The faster the airplane, the more time and space is needed. This is one of the main differences, from the pilot's point of view, between lightly and heavily wing-loaded airplanes. In an ordinary trainer, if you want to be flared out and floating level at, say, 2 feet, you may want to begin the flaring out at, say 20 feet. If you were making the same type of approach and landing in a ship of four times the wing loading—say a transport—you would have to start the flare-out at 80 feet! Let's assume you are flying a trainer, and want to start your flare-out at 20 feet: the question is—how do you know when you are 20 feet off the ground?

The clues by which you judge height have really all been described earlier in this book. In a landing, the altitudes to be judged are very small, but the clues are still the same. There is the horizon. Where does it cut across things? Suppose that at the far end of the airport there was a tree, about 50 feet high. As long as you are higher than the tree, you look down at it, and the horizon will be visible above the treetop. The moment the tree begins to grow into the sky, you know that you are getting lower than the top of the tree. If the tree is about halfway in the sky (the horizon cutting through behind it at half height) you know that you are at about half tree height, and so on down.

There is the perspective of familiar things. For clearness' sake let's suppose an exaggerated case: a five-story factory building adjoining the airport. The perspective of its rows of windows would show you with utmost precision just how high you are. The row of windows that seems level, slanting neither upward nor downward in perspective—*that* is the row at whose height you are flying. Five-story factories don't adjoin airports, but the same principle works on an ordinary hangar and even on an automobile or a parked airplane; you can tell whether you are looking slightly down at something or looking level at it or looking up at it.

Next, there is the way things appear "above" or behind each other. Suppose that, as you are gliding in, you see among other things two men, one closer to you, one farther away. If your perspective of them is such that the far one appears "above" the near one, you are much higher than they are. If your perspective is such that the far one appears "behind" the near one, you are right down at their level. This clue works also on fence posts, bunches of grass, parked automobiles, airport boundary markers, and in water flying, on crests of waves.

All this may seem rather labored. The experienced pilot will say, "I never use any of that stuff." It has been explained here to help the student pilot, not the experienced pilot; but the fact is that even the experienced pilot does use these clues and probably no others. He just calls it sense. Any landing field, even the loneliest, most desolate,

A clue to height: Assuming that these men stand on a flat place, not on a hillside, you know about how high you are above the ground. When the head of the rear man will be level with the head of the front man, your own head will be about 5 feet 6 inches above the ground, and in a small airplane your wheels will be just about to touch the ground. Same clue works with trees, housetops, parked cars, and airplanes, and even with blades of airport grass and with the waves of the bay. It is by such perspectives that we judge our landings.

has an abundance of clues of the kind described. And where none of the three clues described here is available—in those places the pilot's depth perception becomes uncertain.

Desert or even prairie can have this effect until a wagon track furnishes the object of familiar size or a ranch house or fence offers a perspective.

Large empty expanses of water make height judgment very uncertain. Playing around in a seaplane at a couple of hundred feet average height far out over the water a pilot frequently gets a little shock when he passes over a boat—because almost every time he discovers, from the boat's apparent size, that he is much higher or much lower than he thought he was. Thus empty open water is sometimes hard to land on. But if there is even one boat near by, or preferably two boats, one beyond the other; or if there is a pier within a quarter of a mile or so—then you have again the kind of clues your eye needs and can make a fairly certain flare-out and landing.

THE HOLD-OFF

Once leveled off at a couple of feet of altitude, you have the problem of timing your handling of the stick during the "hold-off." Remember we are discussing here an extremely "floating" landing; your task therefore is neither to gain nor to lose altitude. If you pull back too slowly, you will approach the ground and touch pre-

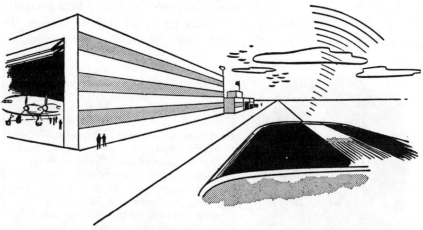

How we judge a landing. Contrary to popular idea, the pilot does not use depth perception in landing. He judges by perspective. An important part of perspective is the horizon, and just where it cuts through things. Here your eye is about 15 feet above the ground. You are raising the nose gradually for a three point landing. . . :

maturely, front wheels first, and you will then probably bounce. If you pull back too fast, you lift your ship away from the ground, and the final stall will then occur so high that you will drop a long way to the ground, resulting in anything from a hard landing to a crack-up. How does the experienced pilot gauge his handling of the stick?

PERSPECTIVE SHIFT

This again turns out to be entirely a matter of knowing what to look for; once you know, it again becomes almost mechanical. Here again, the first requirement is to relax your vision, to take in the whole perspective of the field before you and on both sides. And the thing to watch perspective for is in this case not indications of how high or how low you are, but *indications of whether you are rising or sinking.*

The slightest rise will warn you that you have pulled back too fast, are overcompensating for loss of speed, and are ballooning away from the ground. In that case, you simply wait, hold your hand steady, and let the rise spend itself; only when the ship has sunk back again to the proper level do you resume the backward travel of your hand.

The slightest sink will tell you that you have pulled back too slowly, that you are not compensating enough for loss of speed, and

. . . and here, your eye is about six feet above the ground, which means your wheels are about one foot above the ground. You have raised your nose meanwhile, but do not seem quite in three-point attitude yet. Better get that stick back right now, or you will touch front-wheels first!

that you are about to touch the ground. In that case, you simply increase your back pressure enough to check the descent.

Your indicator of rise and sink—an extremely sensitive indicator—is the perspective before you. Even on a big airport, with hangars, cars, trees, and so on, far away, the slightest sink or rise will produce so marked a shift of perspective of the grass, the runways, of all sorts of small objects before you that you can't fail to recognize it if you have seen it once clearly and have fixed it in your memory. Fortunately, you can fix it in your memory without even using up any flying time; simply look out into an open field, raise yourself on your toes a few times, and let yourself down on your heels and watch how the up-and-down motion of your eye produces changes in your perspective—how all things seem to move into different places behind each other as your point of view moves.

That is the thing to watch for during the hold-off.

THE STALL-DOWN LANDING

The stall-down landing is essentially the same procedure that has just been described, but with this difference: While in the floating landing you glide to almost ground level first, flare out, and start your "hold-off," your final slowing up of the airplane, only after you are almost down. The "stall-down" landing requires that you blend the approach glide, the flare-out, and the slowing up of the airplane all into one maneuver so that, when you arrive at ground level, you arrive in three-point attitude, all slowed up and ready to squat.

This can be done by feel, and if you can make a three-point landing the easy floating way, it will not take much practice to change to the more precise and harder stall-down way. You simply slow up your approach glide when you get within 50 feet or so of the ground and keep on slowing it up. This will make the final float much shorter, and if you start the slowing up of your glide high enough and time the back travel of the stick just right, the final float will become just a brief hesitation on the ship's descent, and there you sit.

Here again it may help to have a clear mental image of just how ship and pilot function during the maneuver

HOW THE SHIP WORKS

The main thing to understand about the ship, the aerodynamics, is the exact manner in which your glide path changes when in a "normal glide" you start pulling your nose up. We can analyze the maneuver best by breaking it up into stages, assuming for clearness' sake a very rough, jerky, but at the same time competent handling of the controls. The normal glide is by definition that glide in which the airplane will perform the shallowest glide, that is, glide the farthest horizontal distance from a given altitude. By generally accepted practice, the normal glide is the glide you use in making your landing approaches. Thus if toward the end of the approach at about 50 feet you start pulling your nose up, you break your normal glide thereby and are bound to get actually a steeper descent. But not immediately.

First of all, you get a temporary excess of lift which causes you to balloon; the glide temporarily flattens, the ship temporarily shoots forward without losing much altitude, and temporarily it seems as if you are going to overshoot the landing spot toward which your approach glide was carrying you. But, after a couple of seconds, the ship has slowed up and the glide path changes to that appropriate for the lesser speed; it becomes much steeper. If that steeper, slower glide is now held undisturbed for another few seconds, the airplane will seem to fall short of the landing spot to which its original glide seemed to be carrying it. Suppose then you apply additional back pressure, holding the stick farther back. Again you get first of all an excess of lift which boosts the ship, flattens out the flight path, makes you shoot forward for a moment almost without sink; temporarily, it will again seem as if you were going to overshoot the spot. But again, after a couple second's time lapse, the ship will slow up, will steepen its descent: it will now descend in a fairly nose-high, very slow glide, close to the stall, and it will descend more steeply than before; and, if held undisturbed in this condition for a few seconds, it will again seem to fall short of the intended landing point.

And so forth: Repeated a few more times, this process makes the airplane descend at steadily increasing Angle of Attack, at steadily decreasing speed, and thus with the nose higher and higher and the flight condition closer and closer to the stall. And it will descend approximately along the line of the original "normal" glide; for at any single stage of the maneuver, the *temporary* ballooning effect of pulling the nose up just about balances the *permanent* glide-steepening effect of having the nose so high. Needless to say, the good pilot will smooth out the various stages of the maneuver into one smooth continuous sequence.

HOW THE PILOT FUNCTIONS

That is how the airplane works during a stall-down landing. Now for the functioning of the pilot. How do you gauge such a landing? The tendency will be to try to watch at the same time your height, your speed, and the attitude of the ship, that is, nose position. That seems reasonable but is useless; nose position especially is not much of an indication. The clue to watch is the intended landing spot and the scenery beyond it and to the sides of it. Once the normal glide has been broken, the process of stalling the airplane down can be gauged entirely by watching the spot and the perspective in which it appears and its apparent motion.

You use essentially the same clues that have been discussed and use them in similar manner. You handle the stick so as to keep the spot coming up at you evenly. Fortunately, here again perspective will register every change in your eye level with surprising clearness. Should the spot fail to keep coming up, you have applied excessive back pressure and are ballooning; should the spot come up fast and move toward the ship's nose, you are sinking fast and need additional back pressure on the stick.

What makes the stall-down landing more difficult than the floating kind is that you might run out of stick travel before you are all the way down; that is, that you might get a complete stall 10 or 15 feet in the air. To do the job successfully you must develop a feel for speed and lift, so that you will sense the reserve of buoyancy that is still in the ship at any one moment. You economize, as it were, on the back travel of the stick so that you "come out just right." When you arrive at the ground, there should be just a last couple of inches of stick travel (and hence, a last bit of reserve lift) left in the ship which you can use—by bringing the stick back rather sharply— as a cushion to stop the descent momentarily and soften the impact of landing.

The advantage of the stall-down landing is that it is more accurate; in a floating landing it is harder to predict just how far the float will carry you into the field. The disadvantage of the stall-down landing is that it is more risky. For example, if you should sink down, all slowed up, into a layer of "dead" hot calm air next to the ground, you would drop right on through; and only a blast of throttle would keep you from pancaking or worse. Such layers of dead air are frequently found—especially if the landing field is surrounded by trees which keep the wind out. Again, on a windy day the air near the ground is choppy, and a gust might throw an almost-stalled airplane into a stall. Many pilots have the idea that, in case of engine failure over rough terrain, they would just slow the airplane up to the limit and mush down perhaps in a partial stall. This is a dangerous idea.

Only an extremely skillful pilot would get away with it, and not even he could do it very often. In a forced landing, a pilot's main worry should be to maintain control and not to stall the airplane. Altogether, then, it seems that the main value of the stall-down landing is its training value. Just because it requires keen sensing of speed and "lift," it develops the student's perceptions.

"WHEEL" LANDINGS

The three-point landing is not the only way to get an airplane down. It is not even the best way. Both air-line pilots and pilots of "hot" service aircraft have long abandoned it. It is essentially an unsafe and unbeautiful maneuver, for it requires that the ship be flown near the stall or actually into a stall, that is, that the pilot throw the airplane deliberately out of control—and near the ground at that. If you want the end of a three-point landing to be brief and business-like, without prolonged floating, you must sometimes slow yourself uncomfortably while still perhaps outside the airport boundary over all sorts of obstructions. If you want to maintain a safe margin of speed, the end of the landing becomes a long-drawn-out float that uses up too much of the landing field—perhaps more landing field than there is. Because ground contact can be made at only one speed—stalling speed or near-stalling speed—the pilot is cramped during his approach and cannot get rid of any excess altitude the common-sense way, that is, by dropping his nose. The whole dilemma becomes the more acute the cleaner and more slippery the airplane is—because, in a clean airplane, the slowing-up process will have to begin so much earlier, and the slightest dropping of the nose during the approach builds up excess speed so rapidly. It becomes more serious also the more heavily wing-loaded the airplane is; because heavy wing loading makes any premature stall so much more serious and recovery from such a stall so much more uncertain. As the pilots of really heavy airplanes like to put it, "Once she is through flying, she is really through."

For these reasons the tail-up "wheel" landing, made at high speed, is getting more attention from many pilots. It is the way air liners and "hot" Army ships are landed, but it is just as easy and just as suitable for the smaller lighter airplane. It is perhaps the natural way of landing an airplane. Certainly it is what the layman would expect to do if you asked him to land an airplane: simply get down and skim on—doing in the up-and-down sense the same thing that he does in the right-and-left sense when he brings his car to a stop alongside the curb.

Just how do you make a wheel landing? If you are very good you can do it by performing an incomplete three-point landing. If

An extreme "wheel" landing. Top: Upon touching the ground the pilot pushes forward quite briskly on the stick; note downward deflected flipper. He thus sets wings at zero or at negative Angle of Attack, killing lift or even creating downward lift as shown in lower left. This kills the tendency to bounce. *Lower right:* The student usually is afraid that airplane might nose over. It is true that the airplane will easily nose over when the tail is lifted too much because center of gravity then moves forward of wheel, but this is when airplane is rolling slowly or standing still. *Lower left:* When rolling fast, the airplane is prevented from nosing over by action of Relative Wind upon stabilizer.

you can time the handling of your stick so that at the moment of ground contact the ship has almost no sinking speed, you will not bounce even though you touch "wheels first" instead of three-point. This is the way a pilot will probably choose when he first wants to try a "wheel" landing. It rarely works; usually he will get a bounce. And that is why this type of landing is considered difficult when actually it is easy.

YOU NEED SPEED

The first thing a ship needs in such a landing is speed—excess speed beyond stalling speed. Speed helps two ways. It gives you quick positive control in all respects, and especially in respect to forward and backward pressures on the stick. And, if the ship flies at fairly high speed, it does not fly so nose-high. This means that the landing wheels are not so far forward of the ship's center of gravity; and that means that ground contact is less likely to produce a bounce.

The second thing a ship needs is a slight shove or forward pressure on the stick, applied at the exact moment when the ground impact effect wants to nose it up. In this respect small ships differ from big ones. Big ones have slower responses, and there is time enough to get the stick forward when you feel the wheels hit the ground. Smaller ships are jumpier; and if you wait until ground contact has actually occurred, you may put that forward pressure on the stick just a quarter of a second too late and out of phase with the bouncing of the ship. You will then get a rough and jumpy landing. In a small ship, there-fore, you yourself must control the moment at which the ship is to make contact with the ground. For that matter, the same thing is frequently desirable in a big ship. Here is how it is done.

PLASTER IT ON

Coming in with a good deal of speed, break your glide so that the airplane shoots along level, half a foot or a foot above the ground. Then, when the spot arrives at which you want to make ground contact, simply push over forward and "plaster" the front wheels on. Then, as you feel the ground, keep right on pressing forward on the stick so as to hold the ship on.

And you need not be afraid of pushing forward on that stick. You won't nose over. This is the most important fact concerning wheel landings, and must be clearly understood. You have no doubt noticed how very easy it is to nose the ship over. You step too hard on the brakes, and over you go. Or, with the throttle well open, you push forward on the stick while your wheels are stuck in sand—and before you know it, you are teetering in the balance, or actually going over. Or, you have lifted the light airplane by the tail. When the tail was low, it was quite heavy; but, when you held it up over your head, it became light; and with outstretched arm, you could hold it in a position where it wanted to go up—and over. Naturally you are afraid of pushing the stick forward in a landing.

The difference is that in all those cases the airplane was standing still or moving slowly. In a "wheel" landing, the airplane is in fast forward motion—rolling on the ground at approximately take-off speed! And in that condition, you probably couldn't nose over even if you rammed the stick all the way forward! You might however dig your propeller tips into the ground—so don't try it. The reason for this peculiar stability is once more the Relative Wind—its direction and strength. If you roll at such a speed with the tail low, it will be positively eager to come up to level attitude, because the Relative Wind blows against the underside of the horizontal tail surface and blows it up. The tail will be willing to go quite a little higher than level attitude if you hold the stick forward, thus holding the flippers down. But, once the tail is up, the Relative Wind blows against the top surface of the stabilizer and blows it *down!* At much lower speed, or while standing still, the same tail-high attitude might be extremely dangerous.

There is no reason then not to put a distinct forward pressure on the stick and force the ship into a slightly nose-down attitude in which its wings can develop no lift or actually develop downward (negative) lift, and bouncing becomes physically impossible.

Nor is there any reason not to get on the brakes quite heavily. Immediately after contact with the ground at a speed, say, 20 m.p.h. faster than stalling speed, most airplanes will skid their tires rather than nose over. Naturally this must be taken with a grain of salt. It depends on the brakes, the tires, the top-heaviness of the airplane when resting on its landing gear, and it depends very much on the runway surface. It is obviously true of wet grass; obviously it would be unwise to try it on dry concrete. Obviously, too, the combined effects of forward stick and heavy breaking might throw an airplane; but any pilot would automatically come back on his stick as he felt his brakes pulling the airplane down by the nose. The point is that, after a fast ground contact—immediately after the ground contact—the airplane is amazingly resistant to nosing over; and this is the time when you can really kill some speed with your brakes. Later, as the speed slackens, at the time when the pilot feels much safer on the brakes, the ship is actually much crankier. Thus, in a wheel landing, you must gradually ease up on the brakes as you slow down.

The high speed of a wheel landing would produce a very long landing run. But the very speed enables you to put the brakes on hard; if well done and done right away, this will cut the landing run down considerably. And there are many other advantages. Because you can touch down at practically any speed, your approach is simplified. You can come in rather fast, even nosing down if necessary to kill altitude. And because of the brisker gliding speed you can come closer to obstructions with less risk. Once across the obstructions, you can cut out all floating and hence much waste of space. Again, you can even nose the ship down to get it down. You can plaster it on at the very beginning of the runway while the slowing-up process of the three-point landing might use up hundreds of feet before the ship could even touch down. Hence the total runway length required to come to a stop will often be less in a wheel landing.

OUR LANDING GEARS ARE WRONG

All these refinements of landing technique are only a sign, however, that our conventional landing gear is really all wrong; all this finely timed, quick juggling of the controls is necessary only because the landing gear is *unstable*. The airplane, which in the air always wants to do the right thing, wants to do positively the wrong thing when making contact with the ground. It *wants* to bounce!

That is one reason why the so-called "tricycle" landing gear is sometimes also called the "stable" landing gear: it solves this particular problem. In this connection, when we are discussing the manner in which the gears behave at the moment of ground contact, we should perhaps call it simply the "level-landing" undercarriage; for in this respect its important feature is not where its wheels are in relation to its center of gravity or how many wheels there are or even which ones are castering free and which ones are constrained; the important feature is simply that the airplane, when at rest on such a gear, is in a level attitude much as in cruising flight. Some even sit in slightly nose-down attitude, as if gliding. It is this feature which makes such a gear stable on ground contact.

With the conventional gear, ground contact slaps the airplane's nose up and its tail down. This puts it into a nose-high attitude, and gives it more Angle of Attack and hence more lift. The airplane "bounces," that is, lifts itself away from the ground. With the level landing gear, the ground contact slaps the airplane's nose down and slaps its tail up. This lowers the airplane's Angle of Attack. Its wings spill their lift; it becomes heavy, and hugs the ground.

Thus a landing on such a gear is essentially a do-nothing proposition. The airplane takes care of itself. The pilot can trust it, rather than having to fight it.

That such a landing requires virtually no action, and hence virtually no skill, is only one advantage of such a gear. Another advantage is that it enormously simplifies the tricky business of approach judgment and glide control. The landing can be made with equal ease at any speed between cruising and stalling; the airplane will in any case immediately assume a no-lift attitude.

This means that if your approach is a little too high, you can simply nose down; or, if it is a little too fast, you can set the airplane down nevertheless where you want it down.

The stability of such a gear also makes it possible, if desired, to land without any flare-out; the pilot can simply let the airplane glide until it hits the ground. On a fast, heavily wing-loaded airplane, the ground contact, though stable, would be brutally hard. Hence on such an airplane the descent must be shallowed out by using some power. But that is proper technique anyway for such airplanes, regardless of the landing gear employed. And the advantage remains that the pilot need not know where the ground is but can simply keep flying until his landing gear takes over. This feature makes such a landing gear especially suitable for night landings and for "blind" landings by instrument indications only.

And then the "level-landing" undercarriage has an additional advantage that will become very important in the near future: it makes it possible to build airplanes that are nonstallable and hence nonspinnable and very safe. The only practical reason for equipping the airplane with controls that enable the pilot to achieve a stall has always been that the three-point landing requires a stall.

On an airplane that need not be stalled for the landing, it becomes possible to restrict the backward travel of the stick so severely that the pilot simply cannot achieve stalling speed or stalling Angle of Attack and hence cannot get into the most serious kind of trouble.

And even that is not all! Once the airplane can be rendered unstallable (which can be done if it has a level-landing undercarriage) it no longer needs a rudder! For, as we have seen, the only really important function of the rudder is the additional control in gives the pilot when the airplane is stalled or spinning and its ailerons are "out." Altogether, then, the level-landing undercarriage simplifies flying enormously.

In that case—why was the conventional nose-high landing gear ever adopted? The Wright brothers did not use it at first, but landed level—on skids. Why hasn't the nose-high landing gear been abandoned long ago? It was adopted largely because it would enable the airplane to touch the ground at the slowest possible speed, that is, close to stalling or actually stalled. This is, however, no longer a good reason to retain this form of landing gear. For one thing, it is quite possible to land an airplane with a level landing gear very slowly. The airplane is simply made to touch down in flight at high Angle of Attack, touching its two main wheels to the ground while the nose wheel is still in the air. For another thing, slowness of landing was more important 20 years ago than it is now. With modern tires, springs, shock absorbers, and brakes, we need no longer be afraid to land our airplanes at rather high speed. Why should the pilot worry about landing at 80 m.p.h. on a broad open field when his wife thinks nothing of driving an automobile down a narrow highway at 80 m.p.h., with only inches to spare on either side? Moreover, the argument that landing speeds must be kept low is largely an argument around a circle. High landing speeds are dangerous largely because the conventional landing gear is unstable—unstable not only upon ground contact but also during the entire landing run itself, as will be shown in the next chapter. The level-landing undercarriage may mean a higher landing speed; but, since it is stable both on ground contact and during the entire run, this higher speed may well be less dangerous.

The best reason why the traditional landing gear is still widely used is not connected with landing at all. The nose-high landing gear gives the best possible performance on the take-off—especially if the field is rough or soft. It is a poor landing gear but an excellent take-off gear.

Chapter 17

THE LANDING RUN

MORE airplanes are wrecked during the landing run than during any other maneuver. Just at the time when the pilot tends to think his worries are all over, the airplane is actually in its most vicious mood. Here is what happens: The airplane swerves a little, say to the left. The pilot, dumb, happy, and relaxed, thinks, "Come on baby, keep straight" and bears down a little on his right rudder. But the swerve fails to stop. On the contrary, it gets sharper. The pilot, now awake, gets busy on his rudder and his brakes, but it is too late; within less than a second, the swerve has become a vicious skid, similar to the type of automobile skid in which the rear end swings clear around. Full right rudder and hard right brake are powerless; on the contrary, the brake seems to make it even worse! And while the airplane thus whips around it also heels over to the right: the left wing rises, the right wing goes down, drags on the ground—and cracks!

And the pilot's face is red—a ground loop! There are indeed situations when an intentional ground loop is good piloting—as a desperate means of stopping almost on the spot. After a forced landing in a short field, it is certainly better to ground-loop the ship than to run head on into a tree. In taxiing about the airport, it is better to ground-loop and crack a wing than to collide with another airplane, wrecking two airplanes and endangering lives. Such an intentional ground loop is brought on simply by kicking hard rudder—and you will be surprised at the quickness of the ship's response; you will suddenly realize that during every landing run you have been sitting on a box of dynamite. Some experienced pilots argue, however, that in such situations it is safer and less expensive to tramp hard on the brakes and let the airplane nose up and go over on its back. On the other hand, there are some airplanes in which a ground loop can be performed, with luck, without capsizing, especially if the ground is a little slippery.

Usually, the ground loop is the result of carelessness—and ignorance. If student pilots understood more clearly what causes a ground loop, they would find it easier to prevent them.

WHAT CAUSES THE GROUND LOOP?

Assume that, in making a landing with a cross wind from the left, you have neglected to compensate sufficiently for the drift. You are thus moving slightly sideways to the right at the moment when you

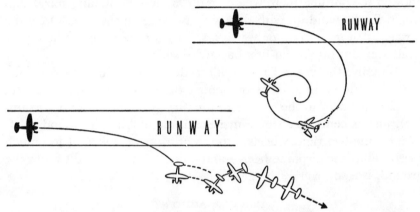

The ground loop. An airplane, running on the ground on conventional landing gear, is *directionally unstable.* Unless the pilot controls it continually and alertly, it will ground-loop. *Top:* If the ground is dry and traction good, it will go into an ever-tightening turn which soon becomes uncontrollable. At some point in the turn, centrifugal force will probably capsize it to the *outside,* so that the wing will scrape and break. *Bottom:* If the ground is slippery (wet grass, ice) the airplane may end up running tail-first.

make contact with the ground. The ship's main wheels immediately try to arrest that sideways motion. But the ship itself—the whole heavy mass of it—once moving sideways wants to continue to move sideways. The ship's center of gravity, however, is located aft of the main wheels. The tail wheel is free to caster and does not guide or constrain the tail; it merely supports it. Thus, as the ship's mass tries to continue its sideways motion and the wheels resist it, the ship's tail swings to the right, its nose to the left.

Unless this tendency is instantly checked by strong, quick opposite (right) rudder the ship's rolling path will therefore swerve to the left. That would not in itself be so serious. The vicious thing is that this swerve will now cause centrifugal force to develop. This force, too, acts in the ship's center of gravity; it, too, pulls toward the right; and again, since the ship's center of gravity is located aft of its wheels, this force now helps pull the tail to the right, thus swinging the nose to the left; the ship's rolling path swerves to the left even more sharply.

The sharper swerve now sets up an even stronger centrifugal force! The stronger centrifugal force again tightens the swerve. This again increases the centrifugal force—and so on. If the process were allowed to continue there would result an ever-tightening spiral that would finally wind up with the ship's twirling on the spot. Or, if the surface is slippery, (ice or wet grass, for instance,) the airplane may straighten out into a tail-first backward slide!

Actually, the ground loop seldom develops so far. For the same centrifugal force that keeps tightening the swerve also tends to heel the airplane over to the outside of the turn (to the right in our example), just as centrifugal force makes an automobile go around a left turn on the two right wheels if too tight a turn is taken at too much speed. Thus the airplane heels—and when the wing tip digs into the ground, it is all over.

TRIGGER ACTION

The thing to understand clearly is this: The ground loop is not caused by the sideways drift, nor is it caused, as students sometimes think, by the pressure of the cross wind on the ship's tail surfaces, causing the ship to weathercock; such things merely cause the first mild swerve. The first swerve might also be caused by some other disturbance—say, careless use of brakes or clumsy footwork on the rudder or uneven ground or a soft spot that retards one wheel. In any case, the first swerve is only the trigger, as it were, which releases the bigger and more vicious effects of centrifugal force.

What really causes the ground loop, then, what makes a slight swerve tighten itself into a catastrophic spiral, is the nature of the conventional landing gear itself: because its wheels are forward of the ship's center of gravity, *any* swerve will always set up forces that will make this swerve worse. The ground loop feeds on itself.

If the tail wheel is steerable and linked up with the rudder pedals, or if its castering mechanism can be locked, the tail can't swing around quite so freely and the ground-looping tendency is much reduced; but even such a tail wheel can't be trusted, since there is not enough weight holding it down—on a rough field, it may lose contact with the ground, or it may skid.

The conventional landing gear, in other words, is *directionally unstable*. It will not keep rolling straight ahead any more than a stick, stood on its point, will keep standing upright. To keep the airplane rolling straight ahead, the pilot must *work*. He must perform with his feet on the rudder and the brakes a sort of balancing trick that is quite similar to the juggling by which you can keep a stick standing up on your finger—a continual series of small corrections that are made so fast as to amount almost to anticipation. Once you let the thing get away from you, it will topple.

To the pilot of a light airplane this may seem rather an overstatement. In the usual light airplane of today, the trick is not at all difficult and is performed by the average pilot without conscious effort. But in an airplane rolling at high speed, the pilot does have to work—and work fast. The centrifugal force of any given swerve increases "as the square of the speed" at which the swerve is taken. Increase the rolling speed from 30 to 60 m.p.h., and the same slight swerve will produce four times the centrifugal force, that is, four times as powerful a tendency to ground loop!

Some light is thrown on this problem by the precautions that automobile speed drivers take to keep their path absolutely straight. At 300 m.p.h. or so the swerve that would be almost imperceptible to the human eye would nevertheless develop sufficient centrifugal force to roll a racing car over sideways! Thus a black line is painted on the salt flats, carefully lined up by high-priced surveyors, to help the drivers keep straight. And one speed driver even uses a sort of gun sight, sighting through it at distant mountains while he drives! In airplane landings, we do not roll at such fantastic speeds; but we roll in a vehicle that is directionally viciously unstable, while the automobile is fairly well behaved.

Thus the rolling path of a really fast-landing airplane must be kept straight with almost superhuman accuracy. This is what limits the landing speeds which the average pilot (even the average military pilot) can handle: it is not so much the length of the landing run

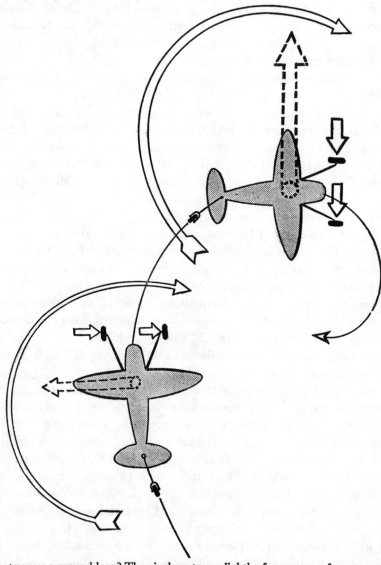

What causes a ground loop? The airplane turns slightly, for any one of many reasons.
First station: The turn causes centrifugal force, pulling on center of gravity, as shown
by straight arrow. Main wheels are constrained by ground, as shown by small arrows;
tail wheel is free to caster. This combination of forces makes the airplane turn more
sharply, as shown by curved arrow. *Second station:* The sharper turn then makes a
bigger centrifugal force; the bigger centrifugal force then makes an even sharper turn:
and so forth.

resulting from faster landings, but the difficulty—almost impossibility —of keeping the airplane straight enough.

OUR LANDING GEAR IS WRONG

Here, again, it turns out that our conventional undercarriage is all wrong—considered as a landing gear. It is excellent take-off gear. What makes the airplane *prefer* to ground-loop is simply the fact that its center of gravity is behind its main wheels. In this respect, too, the tricycle landing gear is the solution. We have seen that, on the conventional landing gear, the airplane would really prefer to roll tail first and that the ground loop is really nothing but the airplane's attempt to get its tail out in front. Well, the tricycle landing gear, "hind side to" as it first appeared to many pilots, takes advantage of that tendency of the airplane. It puts the center of gravity forward of the main wheels, where it "wants" to be during the landing run: the airplane is suddenly rendered stable. On a tricycle landing gear, the airplane "prefers" to run straight ahead. Should some outside force—such as a cross wind or the pilot's clumsy footwork on the brakes—start a swerve, the plane will then not "want" to curl up into a ground loop. The centrifugal force produced by any swerve will tend, not to tighten the swerve, but to counteract the swerving: the airplane will "want" to straighten out.

And here again, this fact is bound to have a deep influence upon the controls. For it rids us of one more situation in which a rudder is badly needed in an airplane. Next to control in the stall, the most important role of the rudder is to keep the airplane straight on take-offs and landings. With the airplane stabilized, *wanting* to keep itself straight, we can get rid of the airplane's most unnecessary and trouble-making control.

Of course, some way must still be provided to steer the airplane on the ground, since in addition to stability you always want to have control. If the airplane were simply stable in the landing run and nothing else, it might still mean that it would roll beautifully straight smack into the administration building. Most airplanes therefore (and for other reasons) retain the rudder. But some of the safety planes solve the problem more brilliantly by making the front wheel steerable and linking it up with the same control which also operates the aileron, that is, with the pilot's stick or control wheel.

When a pilot trained on conventional airplanes is first asked to fly such an airplane he will usually exclaim: "But what about cross-wind landings? Without a rudder, how am I going to land in case of a cross wind?" It may be well to take up briefly the technique of cross-wind landings, both on the conventional airplanes and on the new rudder-less tricycle-geared safety airplanes.

On the conventional airplane, the cross-wind landing requires definite and finely gauged action on the pilot's part. Since the airplane is drifting in the wind, it will touch the ground, should the pilot do nothing about it, while going sideways as well as forward. And because of the instability of the conventional landing gear, this sideswiping sort of ground contact will almost instantly trigger off a ground loop. Hence the pilot must kill the sideways motion just before the ship touches the ground. If you understand wind drift, as explained earlier in this book, you can see that there is only one way of having straight, head-on motion over the ground while flying in a cross wind: the airplane must slice sideways through the air. If the cross wind is from the left, the airplane will tend to move, relative to the ground, sidling toward the right; and the only way to make it go straight relative to the ground is to make it slice leftward through the air.

There are maneuvers which will do this. One method is to dip the left wing slightly down by aileron, at the same time keeping it from making a left turn by putting on some right rudder. This results in a sideslip toward the left. If done to the correct degree, this leftward motion of the airplane *through* the air will just cancel its rightward motion *with* the air and the net result will be straight forward motion relative to the ground. This sideslip is usually blended into the hold-off of the three-point landing. Just before the final touch-down the wings are usually leveled, though it is entirely permissible, in a strong cross wind, to make even the actual ground contact with the windward wheel first.

Another method is to fly the airplane in a normal fashion, allowing for drift in the usual way by "crabbing." In that case, the pilot simply makes sure that his actual flight path is exactly down the intended runway. With a cross wind from the left, this means of course that the **airplane** is *pointed* to the left of the intended landing direction. Just before ground contact, at the very last half second, the pilot then applies rather abrupt right rudder and thus yaws the nose into the direction in which the airplane is actually traveling—that is, straight down the runway. The wings are held level during the yaw maneuver. If you analyze the maneuver you will find that it, too, amounts to a momentary leftward slicing of the airplane through the air.

You notice that both methods require the use of a rudder. Hence the question, in regard to the rudderless safety airplane: "But what about cross-wind landings?"

The answer is that on the stable tricycle landing gear, the airplane can afford to touch while "crabbing," that is, while moving relative to the ground in a sidling fashion. The resulting sideswiping sort of ground contact will simply give the landing gear a chance to show its stability. Consider once more our example, where the cross wind is from the left and the airplane's nose thus points to the left of where the airplane is actually going. The nose wheel, remember, is free to caster, and the ship's center of gravity is ahead of its main wheels. At the moment when the main wheels make their slightly sideswiping contact with the ground, the ship will cock itself sharply around toward the right until its nose points in the direction in which it is actually moving. And it will then move more or less straight ahead, with no tendency to ground-loop. In a strong cross wind, when ground contact is very markedly sideways, this initial cocking around happens instantly and with surprising quickness. In ships whose nose wheel is steerable and connected with the aileron control, the quick castering of the nose wheel is felt by the pilot's hand as a distinct kick of the control wheel. But since this swerve around is a stable reaction rather than an unstable one, an anti-ground-looping swerve rather than a ground-looping one, it makes no demand upon the pilot: the pilot simply lets it happen.

HOW THE FOOLPROOF AIRPLANE FOOLS YOU

Yet it is a fact that experienced pilots have got into trouble on cross-wind landings with rudderless tricycle airplanes. Characteristically, the experienced pilot will sometimes have trouble while the complete novice lands such an airplane successfully and perhaps doesn't even realize that he is landing cross-wind with a lot of drift. Since this type of airplane is becoming more important all the time, the matter is worth explaining.

It appears that in a two-control (rudderless) ship with steerable front wheel there are two phases of cross-wind landing that can be bungled.

Here is what happens (still assuming that the cross wind is from the left): Just as the ship is about to contact the ground, the drift suddenly becomes rather alarming to the pilot, and he tries to kill it in the manner to which he has become accustomed in conventional airplanes: by slightly dropping his left wing. He does so probably whether he intends to do so or not. In practical flying, probably most of our landings are slightly cross-wind; we are almost every time "taking out" a little drift by dropping one wing or another—until that correction becomes quite automatic.

The experienced pilot, then, drops his left wing just before making contact. Since the rudder is linked with the ailerons in those ships, and a crossed-control maneuver such as a sideslip is therefore impossible, this is useless—though it does no particular harm. What does harm is the effect upon the ship's front wheel. Since the front wheel, too, is linked up with the aileron control, holding the left wing low results in setting the front wheel for a swerve to the left—just at the moment when the very idea of the tricycle gear demands that the front wheel must presently be free to caster around to the right!

Two bad results follow. First, the front wheel hits the ground with a nasty sideways swipe, which in extreme cases can lead to damage of the front gear, or, if done often enough, even to its collapse. Second, the front wheel's being angled to the left and being held there by the pilot restrains the ship from executing that all-important swerve around to the right which would line it up with its actual motion, relieve it of all sideways stresses and stabilize its landing run. In short, the very idea of the tricycle gear is nullified.

A WRONG CONNECTION

This is not good, but some pilots pile a really bad mistake on top of it. At the moment when the ship touches the ground with sideways drift to the right, it is inevitable that there will result a tendency for the ship to heel slightly over to the right; that is, the left wing will tend to come up, the right one to go down. This is not in itself serious.

It is but an expression of the ship's general tendency to move toward the right; a result of the very force that will presently swerve it around to the right; and as that swerve takes place, this capsizing tendency will automatically relieve itself! It is much more pronounced if at the moment of ground contact the pilot is holding a little left aileron (in the manner and for the reasons just described), thus holding the front wheel over to the left and not giving the ship its head. In that case, with the ship not free to execute that swerve around, the capsizing tendency cannot relieve itself instantaneously, and the left wing will come up much higher.

Where the experienced pilot now makes his serious mistake is in his reaction to this heeling over of the ship; sensitive as he is to any lateral maladjustment of his ship, he reacts by giving left aileron. He thinks something like, "I've got to keep those wings level," or "I must not allow the wind to get under my left wing and throw me over." But he forgets again that his steering wheel controls not only the ailerons, but the nose wheel as well. In trying to get the left wing down he also cocks the front wheel around so as to steer the ship into a swerve to the left. This swerve to the left sets up a centrifugal force which increases the capsizing tendency; the right wing goes down even more, and the left wing rises alarmingly. To this the pilot reacts somewhat frantically by giving even harder left aileron. He thereby cocks the nose wheel around to the left still more sharply and steers the ship around to the left even more sharply! Centrifugal force increases again, the ship heels over even more, the pilot twists the aileron wheel all the way over, and thus there results something that has all the appearances of a ground loop.

Actually, it isn't a ground loop. Actually, it is caused not by any instability of the landing gear but merely by a wrong connection in the pilot's brain and by his consequent mistaken and cramplike action on the steering wheel. If at any stage during this fake ground loop he would only let the ship have its head, it would immediately straighten out and continue straight ahead; and doing so it would forthwith also level its wings.

PART VI

THE DANGERS OF THE AIR

The dangers of the air are not what most pilots think they are. We are often cocky at the wrong time. And often we are afraid at the wrong time! To tell a pilot that he should play it safe is to tell him nothing: nobody wants to crack up; the question is: just exactly what are the dangers, and how does one deal with them? To tell a pilot that he must keep flying speed is but little more helpful: the question is, just what *is* flying speed in any given situation? And how does one keep it?

Leighton Collins, who answers these questions in the following chapter, is an active pilot of several thousand hours' experience, and is the editor and publisher of the magazine *Air Facts*. Before *Air Facts* became a magazine for pilots, it was a monthly information service to pilots, devoted exclusively to an analysis of air accidents. The following discussion is the essence of six years of *Air Facts'* accident analysis.

Chapter 18

THE DANGERS OF THE AIR

By LEIGHTON COLLINS

HOW much of an art flying is, how difficult it is to work stick and rudder correctly, can be gleaned from understanding, in proper perspective, the cases when the art of flying breaks down and the pilot fails to keep flying and crashes.

Perspective in this is quite important, because it is possible to prove or disprove almost anything you want with aviation safety figures. As far as the air-line pilot and passenger are concerned, flying is a particularly safe form of transportation. As far as the student pilot, both civil and military, is concerned, learning to fly is also a safe venture. Beyond these two categories, however, there are many pitfalls, and rather than being told what the general accident frequency is, an inquiring person is usually presented with the per cent improvement in accident frequency during recent years. Or stress will be put on the fact that only one out of every several hundred accidents involves fatalities, an accident being anything from a scratched wing tip on up. The truth is that our over-all accident frequency, that is, in fatal accidents, is far from satisfactory, and much of the "improvement" indicated in the figures of recent years comes mainly from inclusion of a large volume of especially favorable figures from the training category.

Some people, a pioneering minority, believe safer airplanes are the answer to the safety problem in general civil flying. That such airplanes can be built is no longer open to argument. They have already been built, and in sufficient numbers to develop enough in the way of statistics to prove that their good record is not simply from lack of a large enough exposure. The main obstacle in the way of the safer airplane is not aerodynamics but a quirk of human nature. A part of the zest of flying is its potential danger. A great part of a pilot's pride in his skill is that he is able to fly safely mainly because of this skill. He doesn't want it made safe for him; he wants to make it safe for himself. To some extent he is right about that. A considerable part of flying accidents are not related to lack of piloting technique but to a plain lack of judgment. On the technique part, he is way off though—he simply is not delivering. And no one knows that quite so well as the layman, and the pilot's friends and relatives on the ground.

How is this to be interpreted? Have the training advocates proved anything more than that it is possible, under their particular system of regimentation, to teach people to fly safely while they are learning, that is, in conventional airplanes? Actually it does seem that is all they have proved. Admittedly good while it is going on, the best we have been able to devise in training systems still dumps pilots out into groups turning in a fatal accident in often much less than every 10,000 hours flown, which is ten times worse than their training record, and roughly ten times as hazardous as driving an automobile. The question of why people fly no more safely than they do when they are on their own, free of prescribed flight exercises, free of the watchful eye of their flight instructor, is yet to be answered.

Wherein does piloting technique, in its present stage of development, fail to carry the ball? Let's consider it from the standpoint of what the pilot thinks happens and from the standpoint of what actually goes on.

Here is the way an average pilot thinks of his own flying risks: He worries mostly about an engine failure, feels that if that engine will just keep going he will too. Actually, bona fide forced landings account for about 6 per cent of our fatal accidents each year. Obviously it would be unsound to attempt to minimize the seriousness of a forced landing when it does come, but it is important to point out that we often go a whole year without a fatality in the type of forced landing that the average pilot would regard as typical—a fatality growing out of a landing on rough terrain, with resulting overturning, ground looping, and that sort of thing—or from collisions with objects during the approach. What the pilot does not realize is that usually in a whole year's crop of fatal accidents following a motor failure every one of the airplanes was found with the nose in the ground, tail in the air, spun in. That can only mean that the greatest hazard following motor failure is a loss of control of the airplane growing out of the pilot's misusing the controls in attempting to maneuver excessively and abruptly. An airplane can spin only with the pilot's help. Pilots who sustain a motor failure should instantly regard their then major risk as spinning the airplane. And they should realize that these spins are almost always out of turns—tight and quickly entered ones made in an effort either to get back to the take-off field or to get into position for an emergency landing in a random field. Actually, then, the real hazard following a motor failure is not the forced landing, but the spin. When there isn't any spin, airplanes are landed throughout the year in incredibly small places and are unbelievably damaged in nose-overs, ground loops, and collisions, without fatal injury to the occupants.

Think now, do you remember an air liner "landed" in western Pennsylvania in a forest in which the tree trunks were up to 6 inches in diameter? Do you remember a similar ship landed in an apple orchard between Dallas and Fort Worth? Can you remember the case of the pilot who started up a narrow blind canyon in California which went up faster than he could in his Culver, and who finally "landed," power on, at 7,000 feet elevation, in a rocky, boulder-strewn area? Surely you've seen pictures of light airplanes perched in treetops—into which they went in a glide, or steep climb with full power. Isn't the absence of fatalities in these forced landings at sea in rough water in cargo landplanes and bombers also impressive? Haven't you seen some force-landed airplane hauled back to your own airport all rolled up in a ball with the pilot sitting dejectedly on top of it? The moral of this type of case is this: Airplanes do all right by their occupants if only they are got on the ground under control. Surely, if pilots were less afraid of ground contact, they would behave better during the approach. Recalling successful forced landings, however, is not so easy as recalling the ones in which there were fatalities. But in doing that we are all likely to again miss the main point: In the fatal forced landings the airplanes spun in following a motor failure. They weren't really landings at all!

After motor failure (6 per cent), an average pilot regards weather as his most serious hazard in flying. Here again, the pilot is wrong. Weather is far less important than he thinks it is as a source of accidents, for actually it accounts for only 8 per cent of all fatalities. And, too, the pilot's idea of just how the weather might "get" him is also wrong.

A contact pilot's main fear when he is out in weather is that he will suddenly fly into a condition of no ceiling and no visibility. That, of course, does happen at times, but far more often an accident in this category is a story of the pilot's having pushed on and on under lower and lower ceilings and into less and less visibility. And usually he is over terrain with which he is unfamiliar. He still gets into trouble, but at least it is enlightening to know that this particular hazard turns out to be fundamentally a judgment hazard—failure to turn back sooner, or to get down sooner, or not to start out at all. After a few brushes with weather nearly all pilots come to fear weather, not per se but from the standpoint of whether they will have the strength of character to follow their judgment, accept the inconveniences of delayed departures, delayed arrivals.

There is, however, one technique hazard in weather accidents which it is imperative to see clearly: About half of the weather accidents involve a non-instrument pilot trying to fly on instruments. Someone once said that if you will look at an airplane long enough, sit in it long enough, fool with the controls long enough you will decide you can fly it. That truth is also applicable to flying behind a set of instruments—you just decide after a while that in a pinch you could use them successfully. But it doesn't work out that way, and you find these fellows ending up in a spin in, or a power spiral into the ground, or a structural failure, usually within 3 or 4 minutes after they go on instruments if there is much turbulence.

The trouble that a non-instrument pilot has in trying to fly on instruments is that he does not understand instrument lag and tries to carry over into instrument flying his contact flying habit of thinking primarily in terms of the attitude of the airplane. His natural impulses on these things are all wrong.

For instance, take the air speed alone. A non-instrument pilot going on instruments at a cruising speed of 100 m.p.h. is quite likely to run into this situation: A strong impulse to get away from the ground will cause him to pull up. In a few seconds the air speed is showing 60, and the controls feel uncomfortably light. Owing to instrument lag his true air speed at the moment is probably more nearly 50 than 60, and his nose is way up. In order to get away from the incipient stall that he senses, he pushes the stick forward and holds pressure on it, watches the air speed as it creeps reassuringly up. When it gets to 100 again he eases the forward pressure on the stick and thinks he is going to fly along level at 100 m.p.h. Actually at this moment of 100 m.p.h. his air speed is really behind the times; he is really going maybe 120 and has "leveled off" in a diving attitude. Pretty soon he discovers that, even though he has "leveled off," the air speed isn't going to stay on 100 but has run up to 130 or 140. Suddenly he realizes that he is now really headed for the ground, and he hauls back and keeps hauling back until the air speed has settled back down to 100. This brings on a zoom, and when he "levels off" with the behind-the-times air-speed reading 100 he is quite likely to be almost to the top of a loop, or at least is in a whip stall attitude. That time it really gets away from him, and, of course, he does not know how to make a stall or spin recovery on instruments. Meanwhile there are two other primary instruments that he should have been keeping under control, the uses of which are equally as involved as the air speed and toward which he is also certain to react with equal faultiness. A non-instrument pilot's keeping control of his airplane on instruments is about as likely as a layman's getting a conventional airplane up and down without benefit of previous instruction.

Next to forced landings and weather, the average pilot regards structural failure as his third major flying risk. Actually these accidents are also only in an 8 per cent bracket. They are almost always associated with acrobatic flying and hence should at least not worry the pilot who uses his airplane only for more practical purposes. Characteristically, however, these acrobatics cases involve a pilot who has never had any training at all in that type of flying, who yields to a sudden impulse to show off a little. Such a pilot puts far greater g loads on the airplane than it was designed for. While expert acrobatic pilots have for years put on amazing demonstrations with commercial type airplanes, it is to be remembered that they are experts and that a considerable part of their technique consists of knowing how not to overload the airplane with g loads. Civil type airplanes are designed so as to give as much pay load and gasoline capacity as possible along with enough strength to withstand the bumps in rough air and stresses of hard landings. When enough starch is put into the airplane structure to withstand rough handling in acrobatics you have an airplane that will carry two people, no baggage, not so much gas, and in addition one that requires an engine with extra power just to lug along the increased weight of the structure. In a ship that is not stressed for acrobatic flying—no.

And thus we leave the average pilot—who regards engine failure, weather, and structural failure as constituting nearly all of his flying hazards—and is dead wrong.

These things are, of course, serious hazards in flying, but they are, quantitatively, minor hazards. And even in the extent to which they are hazards, the pilot all too frequently thinks them dangerous for the wrong reasons. With so mistaken an idea of where his dangers really are, it is perhaps no wonder that the pilot is likely to get into trouble.

The story of how and why people do break their necks in airplanes is almost unbelievable, most of all to the layman. You can take the last 1,000 fatal accidents in civil flying and divide them into all sorts of stacks—experience of the pilot, make of airplane, type of accident, and so forth, and you find always the same thing—the pilot lost control of his airplane. The fine points of flying technique do not mean anything to the outsider, but in looking at such an analysis he does see sharply one thing which the pilot doesn't: In many of these cases there is an obvious element of exhibitionism. This element is so strong that the layman is likely to decide that pilots are a curious breed, one with a marked deficiency in common sense, for their fatal accidents seldom occur while the airplane is being put to any normal use.

From the pilot's standpoint, of course, it doesn't look that way. He thinks only in terms of flying technique. After all, isn't it often difficult to get a clean, sharp stall and a spin? Isn't it almost inconceivable that anyone could ever miss the pronounced warnings which every airplane gives in too slow flight? He very definitely does not see exhibitionism as a major hazard in flying, nor does he find it easy to believe the real story on flying hazards even after he is told. Why should it be easy for him to believe? He is already a pilot, well trained and with a Certificate of Competency. But that bears only on convention. Neither his textbooks nor his training have mentioned or brought out in practice the facts of life as to flying hazards, that is, as to what they really are, that is, as to the things that he is going to have to understand and guard against if he is to fly safely where innumerable others out of his same mold have not.

Flying releases something almost uncontrollable in the average pilot. Learning to fly the prescribed patterns in training, learning to make precision approaches and landings, learning coordination of stick and rudder in every detail, learning accuracy and to be ever the master of the machine—well and good; but, once on his own, there surges within a pilot a powerful impulse to break the bonds of every restraint that has followed him into the free air. He wants to throw away the music and to play, play as it has never been done before. He wants to give vent to all the suppressed feelings of his innermost self. That is his reaction to this new medium of expression in speed and infinite freedom of motion.

And that is why you soon find him, by the score, flying over to dive at and zoom above his home and spinning in; that is why you find him trying to climb as steeply as possible after a take-off and spinning in; that is why you find him pulling up into graceful climbing turns and spinning in; that is why you find him anxious to give his passengers a little thrill, a little taste of the wonders of free motion; that is why you find him circling low over his friends to wave and spinning in. Little does he realize that these situations bring on conditions of flight outside the pattern of what his training maneuvers have taught him. He does not know how many people mush into treetops in the pull-out from their dives, how many of them collide with wires, chimneys, trees, radio towers, guy wires, and countless other things they did not see in time in their low flying. He doesn't know how likely he is to kill that first rider to whom he is only trying to be a generous host. And above all, he hasn't the vaguest idea that in nearly 70 per cent of fatal accidents the airplane spun out of a turn, hit the ground with the motor running normally.

The pilot has an idea about this, but it is such a blurred image that it is almost useless. He knows that, in most fatal accidents, it is generally concluded that the pilot stalled the airplane at low altitude. In his mind a stall means loss of control from flying too slowly. He thinks of stalls mainly in terms of the way he practiced them in training— throttled engine, slow glide, finally the stick all the way back with the airplane in a laterally level attitude—or, power on, an extremely steep climb with lots of back pressure on the stick and lots and lots of warning from the airplane. This conception of "stall at low altitude" as the cause of most pilot's troubles, mark well, is associated with more or less straight flight, and it does little more than cause pilots to be quite speed-conscious in gliding in for landings. At least it accomplishes this result: By reason of such circumspection few pilots ever stall an airplane in straight gliding flight.

Here is the central fact in flying safety as far as the general-public-type-of-pilot is concerned: Just before the airplane went out of control it was making a turn, usually with power on.

These "turn" cases (70 per cent of all fatalities) come from almost all the classes of fatal accidents. Note especially, for instance, that almost the entire group of fatalities following motor failure goes into it. There are simply almost no fatalities following motor failures other than from the pilot's spinning the airplane; the usual story is that he was trying to make an abrupt turn to head into some selected landing area.

He gets in bad weather and makes an abrupt turn in cruising flight and spins out; he falls out of those turns over people's houses and golf courses by the dozens; he falls out of those climbing turns after a steep take-off; even in his acrobatics at low altitude, wing-overs, zooms, climbing turns, there is still that common denominator indicating that everything was routine until entry into some last form of turning flight. Yet you ask a pilot how most people get killed in airplanes and he will tell you anything except the fact that most of them lose control of their airplanes in turning flight close to the ground.

To understand fatal flying hazards—at least 70 per cent of them—it is necessary then that we get a new grasp, a completely new understanding of what we do and what an airplane does when we turn it. Could we go a year in which no one lost control of his airplane in a turn we would have a civilian flying record of 100,000 hours per fatality. Our flying record would be as good as our training record. There would be only one or two fatal accidents a month in the whole United States instead of—read your papers.

It is difficult to present a logical, step-by-step analysis of the enigma of the turn. When pilots confront a situation like that they usually seek refuge in the old dodge, "You've just got to know how to fly." Do not assume, however, that some really good and well-trained pilots don't fall out of turns. They do. And when they do and survive, their first question is, "What happened?"

By and large we are no more than day-olds in the air. We have lived so long on the ground that we are fundamentally ground animals. It is inescapable that for some time yet we all take into the air with us the fundamental reflexes that we have learned on the ground. To use one example: It is honestly believed that at least three-fourths of the pilots who have lost control of their airplanes in the air and spun in believed, consciously or unconsciously, that the

Nose position and the level turn. While we like to think we fly mainly by feel, it is only natural for us to use most what we can judge best: attitude. For this reason we soon establish in our minds a correct attitude appearance or nose position for a level turn. There is nothing wrong with this, but as a turn steepens, nose position becomes a more delicate thing, especially when we tend to correct it with rudder.

rudder is used to turn the airplane. This idea is aided and abetted by the fact that to this day no positive straightforward statement is given students concerning the role of the rudder or concerning the manner in which an airplane really does turn. Pilots are harangued hour on end for coordination—just the right amount of rudder as they start banking for the gradual turn into one of their eight patterns, or S across the road patterns, or pylon eight turns. But get him into an emergency, or to the free living and flying moment of wanting to turn abruptly, to circle something of interest on the ground; get him in a situation where he wants to turn as quickly as he can, and a ball-bank indicator will show nine times out of ten that he skids his turn.

It is to be expected that a pilot would have this deep-seated reflex. A rudder on boats has been a commonplace for centuries, and you turn it the way you want to go. Of even more portent, we have the same thing in a car. To turn to the right, say, you turn the wheel to the right and keep it turned as long as you want to keep turning. If you want to turn even shorter, you turn the wheel even more. That sort of idea is deadly in the air. Our training maneuvers are, of course, designed to train that out of a person. As far as the safe execution of the training maneuvers goes, the record shows that they do accomplish this. But it also shows that they do not accomplish it when the pilot flies as he wants to, when he tries to meet the need or feels the urge of a more rapidly entered turn than he has practiced in training. And, of course, the airplane is steered with the rudder—on the ground.

While the rudder-turn complex is the underlying fallacy, it is the things that follow in its wake that finally dump the pilot out of his turn, for misuse of rudder leads to misuse of the other controls—we make a mistake in trying to correct a mistake.

The difficulty of proper control use in turns grows out of the pilot's lack of sense for Angle of Attack when flying in a bank. For instance, nose position on or above or below the horizon is linked with great strength to the pilot's conception of Angle of Attack. In straight, wings-level flight, many a pilot's idea of Angle of Attack runs about like this—nose above horizon, high Angle of Attack, possible stall eventually, potentially dangerous; nose on the horizon, if power is on, O.K.; nose below horizon, power on or off, everything to the good, Angle of Attack low, speed being gained. There is a modicum of indirect truth in these ideas, though actually, of course, we can spin going straight up, flying straight ahead, or going straight down.

The minute a pilot begins to lay over into a turn, however, he carries with him his ideas about nose position, and at the same time he fails to pick up a new conception that he should have. Nose position: If he intends to make a level turn and by having tightened his turn too fast with his elevators finds his nose a little high on the horizon, he is likely to think more of getting that nose where he wants it than he is to think of just how it got there. Obviously, bottom rudder will put it down, and there will be an appearance-correctness in that the nose is then where it should be in a level turn. But, of

A normal too tight turn. Here is a view of what happens when a pilot doesn't misuse his rudder, but tends to use too much elevator in a turn: the nose rises. When it does, pilots are much more inclined to suspect they are using too little rudder rather than too much up elevator, though, of course, in a well-behaved airplane, rudder is never properly related to nose position (elevator) but solely to how much aileron you are using.

course, he has executed a skyward skid in doing that and has induced a rolling tendency as well which he will have to stop. That means that along with his bottom rudder pressure growing out of his rudder-turn and nose-position ideas he will have to cross his controls to hold his bank steady, that he will have the rudder one way and the stick the other. That means not only flying the airplane in a skid, but that the downward-deflected aileron on the low wing will result in the outer portion of that wing's flying at a higher Angle of Attack than the other one; consequently if a stall does come it will stall first. Such as this, a skidded crossed-control turn, is, of course, poor flying technique. A correct pilot would put the nose down the same way he got it up, with his stick, by pushing his stick forward until it was on the horizon. But there is a combination here of the turn-by-rudder and mechanical horizon-nose-position ideas which are instinctive and are likely to rise to the surface above all conditioned reflexes acquired in training, especially when the pilot is under pressure.

The new thing which the pilot does not carry into his turn out of his level-attitude nose-position reactions is this: Angle of Attack is controlled solely by the elevators; the more you tilt your vertical axis over to the horizon the less you are able to judge Angle of Attack— even by a survival of the essentially fallacious straight-and-level flight reference to the horizon.

If a pilot is flying level at cruising speed and pulls his stick back to a given position, he gets, say, a rather violent zoom, which if carried on will wind him up shortly at the top of a loop, or in a whip stall. From the way the nose rises above the horizon he is quite impressed and knows just what is going on. Say he ran up $2g$ in that pull-up. Now let him roll into a 60-degree bank, pull his stick back to the same place, run up $2g$ also. There is nothing in nose position to tell him he has "zoomed." With a few seconds lag, his air speed hasn't even gone down much. But, of course, with centrifugal loading in the turn his stalling speed isn't the 50 m.p.h. he has in his mind but is 70 m.p.h. For practical purposes, a pilot could well think of an abrupt turn as a zoom and accord it the same respect. Or he might visualize it this way: The small flaps we have give quite impressive results in slowing an airplane up and permitting a nose-down attitude with only moderate gain in speed. The whole wing is a flap in a turn, ten, twenty times bigger than the potent little flaps we respect so much in level flight. And it can slow an airplane down, if depressed, as it is in a turn, ten, twenty times as fast as a regular flap can in normal use. When your airplane cruises two or three times its level flight stalling speed, you can see how it is that pilots can spin so quickly out of an abrupt tight turn. Actually there are scores of cases in which the airplane was flying along straight, suddenly entered a steep turn, and by the 90-degree point spun out.

Our rudder-turn, nose-position-on-the-horizon complexes are, however, only a part of the booby trap which the conventional airplane sets for the pilot in a hurried turn. The blow-up comes when he senses that something has gone wrong or is about to.

And note carefully that, at this point, the pilot has not yet "stalled" the airplane. He is simply in a crossed-control, skidded, too-tight

An abnormal too tight turn. The bad thing about putting the nose down with rudder in a too tight turn is that you get an appearance correctness, but otherwise your controls are crossed and, as indicated by the wind sock on the nose, you are skidding. Meanwhile, with even just a slight amount of bottom rudder you can hold the nose down to a level "attitude" and overtighten your turn tremendously without getting any reliable visual indication.

turn with maybe a little tail buffeting starting, or a trace of lateral instability, or an overbanking tendency brought on by a gust, or with the nose tending to go down in spite of increased back pressure on the stick.

From the standpoint of physiology, our eyes, our sense of perspective, our inner ear and sense of balance, our reaction time, we are not very well adapted to handling conventional airplanes in turns. We are uncomfortable under g loading; we have no very satisfactory way of judging our Angle of Attack in the turn; we lack a suitable mechanical analogy in nose position in the turn, even just the act of turning itself, in the air or on the ground, provides some degree of thrill and unnaturalness in perspective.

These shortcomings as air animals account for the instinctive reaction that dumps us out of our turns; the pilot, from his ground sense, feels safer when he is straight up and down and going straight ahead in level flight. This is the stick-of-dynamite idea which an emergency sets off in a turn gone sour; the pilot relates safety not to Angle of Attack but to the attitude of the plane. If he gets scared in a turn he is seized with an overpowering impulse to get back quickly to straight and level flight. After all, you can't walk along with your body tilted 45 degrees to one side. We have known that a long time. We are not going to be able to unlearn it in a day in the air.

Then what happens? Few pilots realize how easy it is and what a tricky thing it is to spin an airplane with ailerons alone in an incipient stall in a turn. And, in his haste to get out of his turn and back to level flight, the pilot's tendency is not only to use lots of aileron but to pull the stick toward the top rear corner as well. This not only increases further an obviously critical Angle of Attack, but putting the aileron down on the low wing increases even further the effective Angle of Attack of the portion of the wing in front of that aileron, often precipitates a sharp clean stall of the outer portion of the low wing. And then the ship falls off. Instead of the wing's coming up and the process of getting out of the turn getting under way, the nose chases the down aileron. It is wholly unexpected, and there is an abundance of evidence that even the most experienced pilots are likely to apply and hold momentarily full aileron when a wing drops and an airplane falls off when they happen to be thinking about anything but a spin—such as simply getting out of a turn.

By this point a pilot's time is running out fast—with his thoughts still on recovering from a turn he suddenly finds himself in a nose-down attitude, motor running merrily, and he is getting a close-up view of the ground he never had before. How many seconds do you think it will be before he realizes he is in a spin?

In getting down to rational thinking about spin-ins, we must reckon with a constitutional weakness we all have in the use of levers.

After a conventional airplane is stalled, its controls function in reverse. Now we know that no airplane will spin unless its controls are held in a certain position; consequently it must be assumed that in every spin-in case this was done. We practice stalls, spins, and recoveries, but necessarily at high altitude and only in straight, wings-level flight; and, of course when we practice we know that it is coming

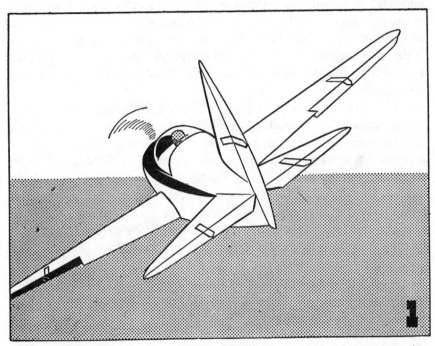

The story of a spin out of a turn. This and the subsequent drawings cover a time lapse of 4 to 8 seconds. Above, the pilot wants to make a quick turn, uses lots of aileron and therefore lots of rudder. A little too much rudder is covered up by his also hurrying the elevators too much, so by the time he gets his bank established everything looks all right as far as nose position goes.

and are all cocked. That this provides some safety, is obvious, but that such practice does not do the job entirely at low altitude is also obvious, else there would not be these spin-ins to explain away.

If you gave a person a car and told him that he could press the brakes hard, but only to a certain point, and that if he went beyond that the wheels would suddenly be cut off their axles, what do you think would happen in an emergency? The pilot knows that if the stick is held back after control is lost the airplane will be locked in a spin from there on, and he has practiced that but with several vital elements missing. When a spin develops unexpectedly, he must have time, maybe 3 or 4 seconds, to do some deducing. First he must recognize the fact of a spin—that takes time; then he must short-circuit his much-used reaction of pulling the stick back to lift the nose and send an impulse to his muscles to push it forward until such time as enough speed has been gained to reestablish the normal function of the controls—that also takes time; meanwhile he is struggling with an instinctive mental and physical paralysis; in practice stalls and spins the ground only says "boo"; in the real thing it comes after you.

In contrast with what a pilot thinks are the principal dangers in flying, namely, motor failure, weather, structural failure, here, then, in brief review is what they really are; here is what really goes on.

Given a need or desire for a quick change in direction of flight, the pilot tends to skid his turn. Under equal pressure of urgency he tends at the same time to try to tighten it faster than it can be tightened without having the nose rise above level turn position. When he sees the nose up, he kicks it down with still more bottom rudder, thereby necessitating further crossing of his controls to keep his bank from increasing. Meanwhile the abrupt tightening of the turn has slowed him down as quickly as a zoom would. At this point he is possibly no more than vulnerable; but, if he does become alarmed, if the airplane suddenly doesn't "feel" just right, he yields to a strong tendency to associate a level flight attitude with safety. Bear in mind that his reaction is not that he is about to stall the airplane. The engine is running, he feels heavy, he knows only that he wants to get out of this immediate situation. His impulsive reaction is this: Wanting to get out of that turn quickly, he moves his stick way over to the high side and at the same time increases his back pressure on the stick. This can cause his low wing to drop precipitately and auto-rotation to begin. But after having dumped himself unexpectedly he doesn't immediately recognize that he is actually in a spin, and under pressure of the ground's being so near and coming up so fast and so strangely he is unable to reverse the reactions that have worked so many thousands of times. He tries even harder to stop the banking (autorotation by now) with his ailerons and still harder to get the nose up the way you usually get it up, by pulling back on the stick. If he survives and can remember anything at all about the flight and you ask him at what point he realized that he had spun he will almost always tell you either "Not at all," or "Just before I hit." And you are likely to get one or the other of those answers regardless of whether it was a half or a three turn spin into the ground. Even

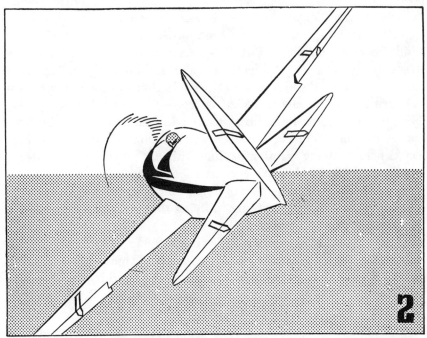

The rudder-turn complex. With his bank established, all that is left with which to hurry the turn is rudder and elevator. The tendency is to think mainly of the rudder, using whatever elevator it takes to keep the nose up. This means having to hold the low wing up also, as excessive bottom rudder is trying to roll the ship to the left. The airplane is now in a crossed-control, too tight, skidded turn.

circumstantial evidence supports this. Autorotation would have stopped at some point on the way down had the controls been neutralized. But it seldom does in the unintentional spin.

Pilots generally are willing to admit that, since an airplane won't spin without certain ministrations on the part of the pilot, our spin-ins indicate a misuse of stick and rudder. But their opinion of their own technique, and knowing eventually personally some of the really good pilots who spin in, prevents a control-misapplication explanation and analysis from satisfying their minds fully. Possibly it should not. As they say, you certainly don't spin out of every bad turn.

Rather than the pilot's stalling his airplane in a sour turn, could there be any extraneous circumstances which might stall it for him? Are there any extraneous circumstances that could trigger off these powerful, smoldering, deadly impulses to misuse the controls when the wholly unanticipated occurs? It seems that there are.

Pilots have argued long and loud on the subject of the effects of wind upon an airplane. But the literature of flying technique is strangely silent on this subject. Considering how rough an airplane can ride on some days, it seems rather odd that so little has been put into print on wind effects. Wind effects—of course, there are none. That is, an airplane flying in a steady wind flies no differently from one in still air. There is, however, unquestionably a large and neglected field in flying training dealing with the subject of gust effects, wind layer effects, thermal current effects, and possibly as yet unnamed others. This is the part of flying that it always seems has been taken for granted, but anything in flying that we take for granted is likely to get us into trouble.

The textbooks are written as if all the flying were to be done in smooth air. As far as it goes, that is possibly all right, but we need to go farther than that. A bump in an airplane is caused by a momentary increase in the Angle of Attack at which the wing is flying. That is, an upward bump, one of those that tends to push your head down between your shoulders. A downward bump, one that makes the belt cut hard across your middle in order to pull you down with the airplane, is the result of a negative Angle of Attack, one in which the wing is not lifting but is pushing the whole airplane down even faster than it would start to fall if it had no wings. We are taught that we control our airplane by keeping Angle of Attack within certain limits, by controlling our lift so as to keep it straight up, or tilted to one side as we do in banking for a turn, so that it will pull us around. But we do not give much consideration to the fact that in rough air Angle of Attack, and hence lift, is a fluctuating quantity, something we can control only moderately. The danger in not being able to control Angle of Attack very accurately from moment to moment is that extraneous causes can result, when we are flying close to the stall, in Angle of Attack getting beyond the stalling angle. That is why attitude alone cannot always be taken to mean the same thing.

Illustrative of this point and also of the extent to which wind conditions can even affect a pilot's opinion of an airplane's flying characteristics is the rather primary situation of flying in a gradient wind. Say the wind velocity at 6 feet above the ground is 5 m.p.h.; at 50

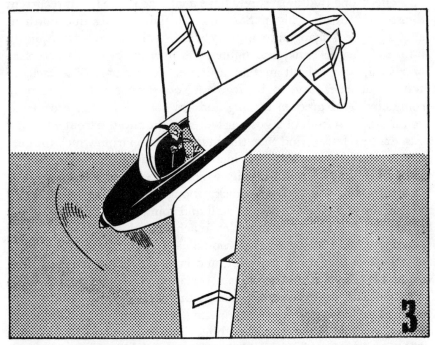

Overboard. While in his crossed-control, too tight, skidded turn, if lateral instability develops and the low wing starts down, or if a gust effect tends to drop it, the most powerful impulse in all flying is to jerk the stick to the right rear corner in an attempt to lift the low wing and the nose at the same time. This is the time of all times and often the last chance to reduce Angle of Attack.

feet, 10 m.p.h.; at 500 feet, 20 m.p.h.; and at 1,000 feet, 30 m.p.h. On the take-off the pilot assumes a normal climbing attitude but suddenly finds the airplane possessed of extra good rate and angle of climb. If the airplane is new to him, he may come back and report that it has exceptionally good climbing characteristics, when as a matter of fact the airplane may actually be below par in these respects on a calm day.

When he comes around for his landing in this same gradient wind these effects work in reverse. As the wind velocity drops that last 5 m.p.h. as he nears the ground he finds himself having to get the tail down in a hurry and decides the airplane has a bad landing characteristic, that it has a tendency to fall out with little warning.

Here, now, is a more advanced wind effect, or, more properly, gust effect, one that must surely have a villainous role in our recurrent disasters in turning flight. Say you are flying along due north at cruising speed, flying into a gusty wind, and you roll over quickly into a 60-degree bank, and tighten the turn up to the 2g necessary to give 1g vertical lift in that attitude. Your Angle of Attack, of course, increases, say, to 10 degrees. Now say right after you have completed 90 degrees of turning and are headed directly cross-wind you fly right in front of a gust, or better, right into the front edge of a gust. Your relative wind then tilts in the direction from which the gust moves, your Angle of Attack is increased considerably, maybe enough to develop that lateral instability which is the first evidence of a stall in a turn, and you become panicky enough to use full aileron to try to level up—and get an aileron stall and spin. Or, in this same cross-wind attitude, say, your top wing reached up into a layer of over-running, faster moving air. This would cause an increase in the Angle of Attack of that wing, more lift, and create a rolling tendency that could be rather abrupt. You can certainly count on it that, if an airplane tends to overbank suddenly, the pilot is going to think first of manhandling that low wing, getting it up with his lateral control device, and he is not going to think first of decreasing his general Angle of Attack to increase his lateral controllability.

Those two illustrations, are, of course, just two of many that could be used to give point to the fact that there are plenty of opportunities for a pilot to approach an incipient stall in turning flight through no very obvious fault or oversight of his own. If you push a person unexpectedly who is standing up, you know exactly what he is going to try to do to regain his balance; jujitsu is based upon such a forthcoming sequence of events. In the air we have equally instinctive reactions when we are overbanked suddenly in a turn, when we "feel" that the airplane is becoming unstable laterally, when there is the buffeting that we have learned to associate with an incipient stall. Unquestionably there are wind effects enough to give us one of these pushes occasionally when we are on the verge of being off balance anyhow, and it is equally true that our rather uniform reaction is to try to scramble back to a wings-level attitude. In that light, our standardized label for falling out of turns, namely, "stall at low altitude," is much too pat. Our trouble is that we stall trying to get out of turns, that we do not know well enough how to make a proper stall recovery in a turn.

The spin. At this point the pilot is bewildered. He is thinking not about a spin but wondering why that left wing won't come up and why the nose won't come up and is using all his strength pulling back on the stick—1 to 4, 4 to 8 seconds, 70 per cent of all fatalities. Only a minority in aviation understand or believe this story. But you fix the controls so that a pilot cannot do this to an airplane (*Ercoupe, Skyfarer*) and no one spins in.

What will put a stop to all this business of pilots spinning in from low altitude in conventional airplanes? Although it has been going on for years, it still does not seem impossible of correction if we look in the right places.

From the pilot's standpoint, and particularly that of the student pilot, we must first of all realize where the trouble is—the turn. That does not mean that we should be afraid of low-altitude turns, but it does mean that we should beware of the desire for abrupt turns. After all, maneuver is the curse of flying—take it out, as in air-line flying, or routine private and cross-country flying—and the accident situation ceases to be a situation.

But, of course, prohibition is never so effective as moderation and education. Striving to improve our turn technique and our understanding of its more somber side will certainly create a bulwark of safety.

Possibly a good way to do that is to make some bad turns and study them, thinking all the while, "Now this is the way people get down low, often in rougher air than this, and spin themselves in."

Get into a 45-degree banked left turn, leading with a little too much rudder as you might if your rudder-turn complex should rise to the surface as it probably would in an emergency. Then use a little more back pressure on the elevators than is proper, as you would also probably do were you in urgent need of hurrying your tuin. This will get the nose too high, so hold the back pressure and pull the nose down with bottom rudder. Notice the amount of bottom rudder you are holding, the *g* load you are under, and that you are having to use top ailerons to keep from overbanking. Now get scared, presumably, and maintain your back pressure and see how quickly you can lift the low wing. Maybe it will and maybe it won't start off to the left; ships vary, and most of the time you need a gust effect to finish off a situation of this kind anyhow. If it doesn't go off, repeat the process, banking a little more steeply, but this time after you get your crossed-control, too-tight turn established, put on just a touch more of bottom rudder. Sooner or later you'll start to fall out of such a turn. What you have demonstrated to yourself up to this point is this: (1) how the rudder-turn complex is a foundation for trouble in getting the controls crossed and permitting more *g* load than is proper even with the nose "on the horizon"; (2) how you can often spin yourself by trying to pick up that low wing; (3) how quickly you can get to stalling speed or stalling Angle of Attack, whichever you prefer or understand best, in a situation of this kind.

And now: Here is the place to really impress upon yourself one of the most fundamental facts of life in the air. Anyone can spin out of a turn—your opportunity to be exceptional is from here on. Just as you get one to start out of the bottom of a turn, make a proper recovery.

A proper recovery from an incipient stall in a turn means simply this: Ease your stick forward a little, reduce Angle of Attack. It doesn't take much. Lateral instability immediately disappears. In fact you've killed not one but two birds with one stone, for you've also lowered your stalling speed by reducing *g* load, and you immediately have a normally controllable airplane. But the main thing to get is this: When you have made such a recovery you should still be in your turn, a looser one, true, but still a turn. If the nose should have got down a little before you moved your stick forward slightly, then you'll be in a nose-down turn, or spiral, but that also is proper. And as you practice these recoveries, look closely at your impulse to think of attitude rather than Angle of Attack, your tendency to move the stick to the high rear corner in an effort to get back to level flight and out of the turn. *That is where people go off the deep end and into the ground.* Think of it not as something you would never do, but as something you are training yourself not to do. Rest assured, it takes a lot of quick thinking and self-control to loosen a turn or nose down when an airplane starts for the ground at 500 feet. Just assume that, if you fly long enough, you'll have one try this, quite unexpectedly, and that you are conditioning your reflexes to make a recovery with no more than 50 feet loss of altitude. You can make such a recovery if only you'll turn the airplane loose momentarily rather than fight it.

Rather than what the pilot might figure out for himself, however, it is much more essential that those who prescribe our training methods should wake up. And in this they should have the moral support of the instructors. After all, we are today teaching flying just as we taught it in the last war. There have been minor refinements in the maneuvers, but they are still basically the same. In 25 years there has not been added to our prescribed curriculum a single maneuver directed at the real killer in flying—loss of control in turning flight close to the ground.

When you talk about crossed-control turns and trying to pull a wing and the nose up when one starts out of such a turn at low altitude, instructors will often answer, sometimes almost venomously, "Possibly so, but we don't teach out students to fly like that." That is just the point—they do. Every one of these pilots who has spun in was a product of the system, the product of a certificated instructor, and he had been checked by a government inspector before getting his certificate. Our only safety effort instructional-wise has been simply to teach a little harder, a little more thoroughly, what we have been teaching all along.

How does the instructor know that his student won't think only of lifting a wing when it starts down at low altitude? Has he made any test of any of his students to see about this? Actually he has, but in only an indirect way, and it speaks for itself that this indirect way is inadequate. And it is even possible that some of those checks inculcate certain misconceptions in the student.

For instance, in the present teaching of control of an airplane in a power stall by rudder, it seems that we are teaching the student to fight the airplane, to manhandle it when he is threatened with lateral instability. Actually, the present rudder-control technique is intended to teach the student not to use aileron at the stall in trying to keep his wings level. That, of course, is sound, though it is seldom stressed as the prime element in the exercise. But why should we teach rudder use as properly related to holding up a stalled wing? Why shouldn't we go to the root of the trouble and teach forcefully that, at the first trace of lateral instability, flying straight or turning, the thing to do is to reduce Angle of Attack. The rudder system puts too much emphasis on attitude as having something to do with safety. Even though a student does get a wing up with rudder, he is still in hot water, he is still on edge, and the danger is that he will think he has learned how to force the airplane to fly when it doesn't want to fly.

The present accent on reverse rudder to stop autorotation in a spin is a further extension of the danger of getting rudder use ahead of stick in the student's mind. Actually some of the trainers have been required to be designed so that opposite rudder in a spin will stop the spin, and the airplane will fly right out with the stick held full back, though usually with much too much g load on the machine, that is, with too fast a pull-out. The significant thing is that the turbulence situation up high is quite different from that close to the ground, where the motion of the air over irregular terrain, buildings, trees, causes a much more scrambled structure in the air. In such turbulence, it is likely that the stalling Angle of Attack is much lower than in high air, and it is both possible and probable that some of these airplanes that will fly out of a spin with opposite rudder up high will not do so down low. And, by the same token, keeping the ship level laterally with the rudder in a power stall may not work out either at low altitude on many days.

There is, of course, a place where teaching a student to augment lateral control by rudder use is sound, namely, in landing and taking off—where he is too close to the ground to nose down any. But that is as far as such teaching should go. It certainly should not be impressed on the student as a means of keeping the airplane level laterally in an incipient stall when he is high enough to spin in. If he is that high, he has room to reduce Angle of Attack, lose a few feet altitude, and gain flying speed. With more speed, he can then maintain level flight at a lower, safer Angle of Attack.

Possibly another weak point in our teaching today is in what little we do teach about control misapplication in turns. Many an instructor has been embarrassed by having a student ask him to demonstrate a stall and spin out of a level turn. The instructor might smoothly enter a 70-degree banked turn, keep off the rudder once in it, keep the g load high, and finally wind up with the stick all the way back, the airplane shuddering, and beginning to go into a turning power slip from which it finally gets into a power spiral, after a lot of turns. These demonstrations have often gone off at a tangent with the instructor showing the dangers of top rudder in a turn producing a spin over the top. Actually, today, that is about the only emphasis put on spins out of turns. It would seem to be a misplaced emphasis, because in the accidents there are 99 spins out of the bottom of turns to one over the top. Maybe they start as often over the top as they do out of the bottom; but, if so, they are also stopped in great quantities. That would be favored by the fact that, as the student went by level attitude on the way over the top, all his natural impulses to get the nose below the horizon would function, whereas they don't function often enough on the way out the bottom.

One further misconception that students seem to be getting from present training is that the tight turn is the safe turn. Their understanding of falling out of a turn seems to connect itself with slipping out of a turn. It is not clear just what maneuver gives them this idea, but of course they have it just backward. Possibly this misconception explains their tendency to want to keep a turn tight when they've been scared and want to get out of it and back to level. Centrifugal loading in the turn, of course, means the Angle of Attack is not only high, but often on the way up. They should have the idea that, as far as safety goes, the tight turn is the dangerous one, the loose turn the safe one. There is no important hazard in a slipping turn, for slipping in the turn is not related to a too high Angle of Attack and therefore lateral instability and stalling. Slipping develops only gradually; the turn becomes a descending spiral; speed is picked up; lateral instability caused by flying at too high an Angle of Attack does not develop. In the slipping turn the pilot can shallow his bank and lift his nose, if it is down, with safety—but he cannot do this safely out of the stalled turn which he tried to keep tight.

The vital thing, however, is not refinement of present training exercises, but that we should add some new maneuvers to the curriculum. Possibly they should be called *safety exercises*. They would not be designed to a certain standard of perfection in execution in the sense that our present precision maneuvers are; they would be designed to show whether the student had acquired certain fundamental conceptions and reactions. This would mean exploring a completely new field in flying training, with the sole objective of curing the unintentional spin out of turns at low altitude. It should be a fruitful research, for we know what the airplane will do under any given set of circumstances, and by now we should know what all too many pilots do under stress.

Two things particularly suggest themselves: First, there should be at least one exercise to teach the danger of excessive maneuvering in slow flight. With the airplane in a steady, steep climb the throttle would be closed, and the student would be asked to reverse his direction of flight (1) as quickly as possible and (2) with as little loss of altitude as possible. In starting the nose down and a turn at the same time, students often get along all right until they have turned 180 degrees and started to level up. Actually they are often entirely happy up to that point. Much to their surprise, however, trying to lift that wing in many ships will start a spin against the down aileron.

At other times, they get the nose down farther than necessary in the early stages and get buffeting or an actual stall in the pull out. In such a maneuver as this the student would be judged by his quickness in loosening his turn. If he were getting an aileron stall, he would learn that from loosening his turn he could not only do away with the lateral instability but could increase his lateral control enough to go ahead and lift the wing. He would also learn a correct recovery in a turn, that is, that the turn continues but at a slower rate. And, if he had to make a recovery starting from a slightly nose-down attitude at some point in the turn, he would learn that such a recovery would be completed with the nose down a little farther than when he started the recovery—not completed when he got it back up to the horizon. The instructor's attention throughout the maneuver would be concentrated on the student's quickness in loosening his turn at the first evidence of questionable controllability. To get the most out of this, a flexible standard depending on turbulence conditions might finally be set for each airplane—180 degrees in so many seconds and with no more than so many feet loss of altitude. And then the student would be reminded that, close to the ground in turbulent air, these minimums would probably double. But, even if they did he'd have this backlog of safety: He could be depended upon not to try to force the airplane beyond what it was capable of as distinguished from what he would like it to do attitude-wise from moment to moment. Then the maneuver should be repeated with cruising power throughout.

The second most needed thing seems to be some way to know definitely and finally that a student will (1) react instantly to unsatisfactory aileron response by reducing Angle of Attack, and (2) that he will not be afraid to nose down when close to the ground, that is, within the customary spin-in altitudes of from 50 to 500 feet. This would get at their tendency to try to force wings to come up when they are within reach of an aileron stall and before they realize that a general stall is imminent. And it would also get at their low-altitude ground shyness following unintentional stalls.

While there may be some maneuver that would teach this, it seems that the best solution would be a mechanical one—a spoiler near each wing tip which the instructor could operate. The test then would be to see the student's reaction when he began to run out of aileron control—whether he would loosen his turn in order to get it or try to bend the stick over farther. A particular advantage of this device would be that it could be used with safety much closer to the ground than we can practice stalls, for the effect could be produced at well above stalling speeds, and releasing the spoiler would reestablish the status quo instantly. There would also be the further and very important advantage that at altitude it would be possible to give the student some really unanticipated starts into spins out of turns, and that's what counts: In real life (as against mere practice) you don't expect a stall.

Briefly, then, here is the bulk of the whole question of flying safety: Under pressure the best of us sometimes make bad turns. Nearly all of us show deadly impulses in trying to get back hastily to a level attitude when a turn goes sour. We are not taught proper recovery procedure in stalled turns. We are ground-shy when our airplanes stall at low altitude; and, even worse, we usually fail to recognize a spin at low altitude at all. There has been no real advance in flying training exercises in 25 years. There is no single training maneuver today which bears effectively on the one trouble which kills pilots.

Only after an unbiased study and rational approach to this problem will it be possible to teach people to fly conventional airplanes with any reasonable degree of safety.

PART VII

SOME MORE AIR SENSE

There are many advanced phases of the art of flying which a pilot may well chose to disregard. "Blind" flying, for example: A pilot may simply decide that it is not for him, that he will not have the necessary instruments, that he does not want to spend the time and money to learn this particular skill and that anyway he does not want to take the risks. Again, navigation by radio might be such a phase, or celestial navigation. Or, very much so, acrobatic flying; a pilot might simply decide that he does not want to fly upside down—and perhaps he will be well advised so to decide.

There is one phase of flying, however, which is rather advanced, and is now studied, under some name such as "Scientific Cruising Control," mostly by airline pilots; and which is yet inescapably important even for the private flier in his light airplane: the effects of speed and altitude on the airplane's performance. Whatever flight a pilot makes, even if he is only joy riding, he is always going to fly at *some* speed and at *some* altitude, whether he consciously chooses them or whether they just happen; and his speed and his altitude are going to have important effects: on the performance of his ship in all maneuvers, on the miles per gallon he gets, the time he makes, on the number of hours he can run his engine before it will need an overhaul, and sometimes also on his safety. The pilot who understands the principles of scientific cruising control will have a much surer touch.

Chapter 19

THE WORKING SPEEDS OF AN AIRPLANE

A THEORIST and a practician were 'way out over the ocean. Suddenly they discovered that they were very low on fuel.

"Gee whiz," said the practician, "what do I do now?"

"Heck," said the theorist, "don't you know? Just fly at the Angle of Attack where

$$\frac{C_D + \left(\dfrac{1.28a}{S}\right)}{C_L} \text{ is least!"}$$

Needless to say, they fell into the soup.

What put them there was the fact that theorist and practician don't speak the same language. This is one of the weakest points in the whole fabric of our aviation. It keeps pilots arguing in the airport cafés over things to which the engineers and the professors have known the answer for 20 years. It keeps pilots cracking up simply because they don't half understand their airplanes. Worst of all, it keeps instruction less efficient than it might be.

And it is mainly because most pilots will not even try to understand any mathematical statement of any air facts. Yet, unfortunately, mathematical statements are just about the only form in which much valuable information is available. The designers and the professors contribute to the trouble by their refusal to express in words what to them seems more conveniently expressed in graphs and mathematical symbols. You might think that, after designing an airplane, the designer would want to give the consumer a set of intelligible "Directions for Use." But that isn't done. The designer and the pilot live in two different worlds.

This is where this book tries to be useful. It assumes that, if there is anything useful to be known about the airplane, most pilots want to know it. And it assumes that an approximate and rough theory of flight in the pilot's head is worth more than a highly refined theory of flight that remains on paper.

HOW TO GET THE MOST MILES PER GALLON

All right then: What should that pilot have done? He should have slowed up, retrimmed his ship, throttled back, and proceeded in nose-high flight at a speed about 5 m.p.h. faster than the speed of his normal glide: at that speed, he would have made the most miles for each gallon of fuel.

This seems a fact worth knowing. Sooner or later, every pilot will be short of fuel and a long way from home; on that occasion all his intuitions will argue him into speeding up, or into stolidly maintaining cruising speed. His intuition will be wrong. If he wants to get home he can't rely on "feel"; he has to know.

But this one is not the only speed fact worth knowing. Being short on gas and a long way from home is only an example. There are many problems in practical flying to which the answer is simply a certain speed. And the speed of normal glide is only one of several speeds that are of interest to the pilot.

Some day, for example, you may be short of fuel, but it may be that you want not distance but simply endurance. Suppose, for example, that you arrive over an unlighted airport and decide to postpone the landing until the moon comes over the mountain. Or, less adventurous, suppose that on a photographic flight you want to spend as much time over your objective as you can spend and still get home on the gasoline you have. You will want to know the exact speed at which your airplane can maintain flight with the least consumption of gasoline per hour.

And so it goes: You want to get the fastest climb—that is, the most gain of altitude in the least time—and the answer is to fly at a certain speed. Some other time you want to get the steepest climb, that is, the most gain of altitude in the least distance—and the answer is again a certain speed. There is a certain speed that will let you cover the greatest distance in a glide from a given altitude; there is an altogether different speed at which in a glide from a given altitude you can stay in the air the longest time.

Some day in rough air, when you see your wings flexing up and down, you will be glad to know that there is a certain speed which will limit the possible stresses on the ship's structure to a value which your ship can surely stand. Some day you will be in a hurry; but you should know that there is a speed beyond which any throttle-pushing is merely a useless waste of fuel and engine life.

In short: Between an airplane's top speed and its stalling speed there lies a regular scale of what one might call *working speeds*, of which each is the answer—the only correct answer—to some problem of

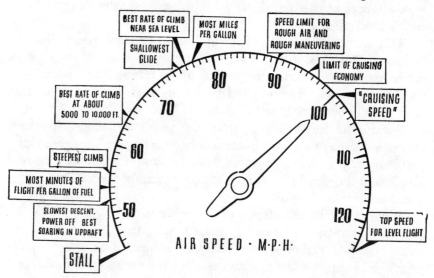

The working speeds of a 100 m.p.h. airplane of average design. For faster or slower airplanes, the figures are different but the proportions the same. For airplanes of extreme design—patrol bombers, fighters, stratosphere ships—the proportions are different.

practical flying. It is almost as definite as if your airplane had a gear box and a gear-shift lever and definite notches for definite purposes.

The diagram shows the sequence at which these various working speeds occur in the average airplane as it is speeded up from stalling speed to top speed.

THE SPEED OF BEST DISTANCE

Let's discuss first of all the Speed of Best Distance because that one is familiar to every pilot; it is none other than the "normal glide." Every pilot knows that, from a given altitude, you can glide the greatest distance by gliding the airplane fairly briskly and not trying to hold the nose up too high. Why is that? The answer is simply that the drag of the airplane is less at an intermediate speed than at either high speed or low speed. Thus, if we were interested in theoretical causes rather than practical results, we could call this speed simply the Speed of Least Drag.

Here is how it comes about. If you fly the airplane too fast, the parasite drag, that is, the drag of fuselage, landing gear, windshield, struts, wires, cooling fins, the skin of the wings, and so on, becomes prohibitively high; it increases as the square of the speed, which is a fancy way of saying that the resistance of the air grows very much if your speed grows even only a little; until finally in a "terminal velocity" dive the airplane can't go any faster even though pointed straight down (or slightly inverted in a no-lift dive), because the drag equals the weight of the ship. Applied to the ordinary glide, this means that, if you try to glide too fast, the parasite drag forces you to point the nose down disproportionately steeply and to sacrifice altitude.

If, on the other hand, you try to fly at a speed slower than the Speed of Best Distance (normal glide), the drag of the airplane also increases. For, in addition to the parasite drag—which is a type of drag the airplane has in common with boats, bicycles, cars, and airships—the airplane is held back also by a drag of an altogether different kind, a kind that is peculiar to the airplane wing and is not experienced by those other conveyances. This is the induced drag— the drag which the wings develop as part of their lifting action. The induced drag is very high when the airplane is flown slowly, in "mushing" flight, with the wings held at high Angle of Attack, the stick well back. The induced drag decreases rapidly as the airplane is speeded up; it practically vanishes in high-speed flight when the wings are skipping through the air at extremely shallow Angle of Attack, and the stick is fairly far forward. Thus the Speed of Best Distance is a compromise speed, midway between too high a speed where parasite drag is prohibitive, and too low a speed where induced drag is prohibitive.

Now, seeing that the Speed of Best Distance is one speed which in power-off flight will get the most distance out of a given altitude, it stands to reason that it must also be the speed which in power-on flight will get you the most distance for your tankful of fuel. As a problem of physics, the two cases are alike; there is a store of potential energy—altitude in one case, gasoline in the other case. And if the Speed of Best Distance is the most effective way to expend altitude, then it must be also the best way of expending gasoline. That's why the pilot out over the ocean should have throttled back and proceeded at the speed of his normal glide—plus a few miles per hour in addition, perhaps 5 to 10 per cent faster than the normal glide.

Why is that? In the glide, with engine and propeller windmilling, the power plant is a drag; in powered flight, it pulls. Then in the glide, the drag of the power plant is among those factors that make it advisable to fly slower rather than faster. With power on, the factor disappears, thus making it possible to fly a little faster without undue penalties. But there are more reasons: When the airplane is flown too slowly, the power plant becomes inefficient; the propeller is designed for best efficiency at higher speeds; the mixture may be slightly wrong, and the engine will burn its mixture less efficiently. Thus, if you fly with power on a little faster than you glide with power off, you do indeed operate your airplane at a less economical speed; but you operate your motor and propeller at a more economical speed, and the net result of all this is a gain in economy.

HOW TO STRETCH YOUR FUEL AGAINST A WIND

And what about wind effects? Head winds and tail winds influence the Speed of Best Distance surprisingly little. If you are trying to make your tankful of fuel take you as far as possible, and you have a head wind, it does pay to add a few miles per hour. If you have a tail wind, it does pay to decrease your air speed by a few miles per hour. The calculations by which you might find the exactly right air speed are exceedingly intricate; moreover, they involve such hard-to-get data as, for example, the exact manner in which your power plant's efficiency changes as the air speed changes. But if you work it out you will find that *wind influences the Speed of Best Distance very much less than you might think.*

If your Speed of Best Distance in still air is about 75 m.p.h., your Speed of Best Distance against a 10 m.p.h. wind is perhaps 77 m.p.h. Against a 20 m.p.h. wind, it is still only about 80 m.p.h. With the same 20-mile wind on your tail, your Speed of Best Distance will still be 73 m.p.h.! In short, it would almost be best to say simply that wind has practically no effect on the Speed of Best Distance. This will be important to remember. For, when trying to make distance, especially against a head wind, one always experiences an itch to fly faster, and then still a little faster. If he wants to get home on the fuel he has, the pilot must guard against a tendency to overcorrect for head wind, and to fly too fast.

THE SPEED OF BEST DURATION

An airplane will maintain flight with the least amount of power if it is flown quite slowly, very nose-high, with its wings at very high Angle of Attack. If we were more interested in causes than in results, we would call this the Speed of Least Power Required; as it is we call it the Speed of Best Duration, for in this condition the fuel consumption is least per hour. The airplane is not covering distance very effectively, for it is too slow; but it develops perfectly astounding endurance. The pilot who is waiting for a ground fog to burn off; the patrol pilot or photographic pilot who wants to hover as long as possible over some objective; and also the operator who sells Sunday afternoon rides by the minute rather than by the mile—they all should be interested in this speed. (In some ships, engine cooling is not very effective in this condition of flight; and in some other respects, too, the engine may not like that very slow speed too well; that sort of consideration is, of course, a different question altogether.)

With power off, somewhat the same is true, except that the ship should be flown even more slowly. An airplane in a glide will make the *slowest descent* if it is glided extremely slowly, in the case of a long-span airplane, such as a glider, almost at the stall. It will in that condition not perform a long glide, long in terms of distance; on the contrary, the angle of descent will be quite steep. But it will perform a long glide in terms of time; it will remain air-borne for a longer time than it would if glided at any other speed. Hence in a glide, too, this is the Speed of Best Duration.

Soaring pilots know this speed and make constant use of it because of its slowest-descent feature. A glider maintains altitude, or gains altitude, by gliding down slowly within a mass of air that is rising; once the glider has been maneuvered into a rising mass of air, its soaring performance depends no longer on angle of glide, but entirely on rate of sink; hence soaring pilots do most of their maneuvering, including steeply banked spiraling, at a speed just clear of the stall.

The pilot of powered airplanes makes use of the same glide every time he slows himself up in his approach in order to steepen his path of descent. What makes it possible to steepen one's descent in this fashion is really two effects—one, that the glide path itself is necessarily steeper because of the ship's getting farther away from the Speed of Best Distance; and two, that the slowness of descent of such a mushy glide lengthens the time it will take you to reach the ground and thus gives the wind more time to set you back. Gliding against a 20 m.p.h. wind, a lengthening of your glide by only 10 seconds will from this effect alone bring you down some 250 feet shorter!

Perhaps the most important use of the Glide of Slowest Descent (or of Best Duration) is in connection with forced landings. One probably should not even discuss this without first giving once more an important warning to any inexperienced pilot who may read it: The great danger in forced landing lies not in hitting some minor obstruction, such as a fence or a ditch, and damaging the ship; the big danger lies in the tendency to maneuver the ship around excessively in an attempt to make a perfect landing—because that sort of maneuvering may lead to a spin and a fatal accident. With that firmly in mind, let us put the question: Suppose you were fairly low with the engine dead and wanted to make a 180-degree turn with the least possible loss of altitude, perhaps in order to head the ship into the wind for a landing; what would you do? Dive it and bank steeply? Keep your nose up and bank gently? Rudder around, and skid?

It can be shown that the ship will come around with the least loss of altitude if it is flown at the Speed of Best Duration (that is, just nicely clear of the stall) and banked at 45 degrees. (In trying this, remember that, in a 45-degree banked turn, the stalling speed of any airplane increases by some 20 per cent over the straight-flight stalling speed; the Speed of Best Duration or Slowest Descent increases correspondingly.)

SOME PRACTICAL THEORY

In some respects, the Speed of Best Duration is the most puzzling of all the airplane's working speeds. Several questions come up in connection with it.

First of all: Why that difference between the power-on and the power-off condition? Why is it that, with power on, it pays to fly a

little faster than with power off? The answer is the same that accounted for the difference between the power-on and the power-off condition in the case of the Speed of Best Distance; it is true that, at the slightly higher air speed for power-on flight, the airplane itself (wings and fuselage) is working under slightly less favorable conditions; but the power plant (engine and propeller) is operating under more favorable conditions and uses its fuel more efficiently; hence there is an over-all gain in efficiency by flying a little faster with power on.

The next question is: How can it be true that the airplane requires the least power when it is flown quite near the stall—when everybody knows that an airplane in such slow flight has rather high drag and that an airplane develops the least drag in flight at intermediate speed, about halfway between stalling and cruising?

Perhaps the simplest answer to that one is: Well, go up and try it for yourself. If you do, you will find it is so; quite near the stall, you maintain level flight at little more than half your cruising r.p.m. If you get too close to the stall, you will find that you need more power in order to maintain altitude. If, on the other hand, you lower the nose and try to fly faster, you will again need additional power in order to maintain altitude.

WHAT IS POWER? WHAT IS FORCE?

Still, the question deserves a real answer. The puzzle is solved the moment you go back to elementary physics and remember the difference between power and force. In everyday speech, we use the two words almost as if they meant the same thing; but they don't. "Force," you might say, is "pull"; it can best be expressed in pounds. The drag of the airplane is a force, and so is the "thrust" of the propeller. "Power" is more complicated; it is the ability to exert a force through a distance in a given time, in other words, to exert a pull of so many pounds at a speed of so many miles per hour. Any child, for instance, can pull an automobile that's stuck in the mud— if you give him the necessary pulley arrangement; he will then exert his weak forces simply for a long time, and the automobile will slowly move out of the mud. But it takes a crew of powerful men to do the same pulling job fast. What the engine of an airplane turns out, what you regulate with your throttle is power.

Hence it is perfectly true that in slow flight, the drag of the airplane being high, the propeller has to exert a hefty pull; but it exerts that pull at so much slower a *pace* that flight requires actually less power—much as you might find it easier to walk up a mountain carrying a pack than to run up a mountain carrying nothing.

Finally, another question is almost invariably asked: I can see, someone says, how all this applies to flight with power on; but what is the connection between all this and the glide? There are two ways of answering that one. The first is theoretical: As a problem in physics, the two cases, glide without engine and level flight on the engine, are identical. In each case, there is a store of potential energy—altitude in one case, fuel in the other. In each case, the problem is to expend this energy as sparingly as possible. The burning up of gasoline or descent from altitude—those are only two different ways of putting out power. If at a certain speed and Angle of Attack the least of one is needed to maintain level flight, then that same speed and Angle of Attack should require the least power output also in a glide.

But another answer will probably mean more to the pilot: Just look at the two flight paths—the normal glide and the very slow glide, and then look at the speed at which the ship proceeds along each path. The pilot who glides slowly will get a very steep flight path, but because of his very slow progress along that flight path he will still be in the air when the faster-gliding fellow, with his shallower glide path, is already on the ground.

MANEUVERING SPEED

Maneuvering speed is that speed, or rather slowness, which will protect your wings from structural failure in rough air and in acrobatic maneuvers. When an airplane flies into an updraft, it receives an upward jolt which is felt by its occupants as a bump. When an airplane executes a sharp maneuver—such as a steep turn, a pull-up into a loop, a snap roll, or the pull-out from a dive, it experiences centrifugal force which is felt by its occupants as heaviness and often referred to as "*g* load." Either case puts a strain on the wings. Every student knows that feeling of heaviness. But it is important for him to realize that just as he himself feels heavy, so the whole airplane actually becomes heavy; that under the influence of bumps and centrifugal forces, the ship is bearing down heavily on its wings. A hard bump or a sharp maneuver adds to the weight which the wings have to support, just as truly as if a heavy overload of sandbags had suddenly been dumped into the ship.

The idea of a maneuvering speed is simply this—to fly so slowly that, when the ship begins to bear down on its wings too heavily, the wings will not attempt to support that additional weight but will stall instead, thus relieving themselves.

In most airplanes, maneuvering speed is arbitrarily set at twice the ships' normal straight-flight stalling speed. If you fly at that speed, you know that, whenever the ship begins to bear down on its wings with more than four times its normal weight, the wings will stall.

Most small airplanes are built to stand up structurally under g loads up to six, that is, the wings are six times as strong as they need be for flight with normal load in smooth air. Hence, if you fly at twice normal stalling speed, you know that *your wings will stall rather than break*.

If you have any reason to suspect that your structure may have been weakened, you can, of course, set your maneuvering speed much lower. If you fly a few miles an hour above stalling speed, the stresses on the ship's structure cannot possibly exceed those of straight flight in smooth air by very much—no matter how rough the air or how rough your handling of the controls. If, on the other hand, you fly at three times your normal stalling speed—which is about top speed in most airplanes—and get into really rough air or start getting really rough on the controls, your structure can fail. For, at that speed, your wings will not stall until the weight on them becomes nine times normal. And, since your wing structure is built to support only six times the normal weight, *your wings will break rather than stall*.

This is an important fact to keep in mind. Flying near top speed in extremely rough air (thunderstorms) the average airplane is capable of breaking itself up; it should be slowed up, and certainly extreme care must be taken not to let it pick up excess speed by diving. Flying near top speed even in smooth air, the pilot has it in his power to tear the wings off his airplane simply by bringing the stick back abruptly. He must handle the stick gingerly, and that is especially important if he should ever be in a dive near the ground—because in that situation, he will naturally be hasty and hence rough. And a dangerous combination is rough air that hits at a moment when the airplane is engaged in a high-speed, highly stressed maneuver.* Moral: If you want to get rough, slow down to twice your stalling speed or less.

LIMIT OF CRUISING ECONOMY

The speed marked *limit of cruising economy* is the speed beyond which it simply does not pay to push the airplane. Strictly speaking, that is of course a statement which the aeronautical engineers cannot make. Whether or not it is economical to use up a lot more fuel and engine life in order to get a little additional speed depends entirely on how valuable your time is—to you! In a race, for instance, or in combat, 1 m.p.h. might make all the difference. But the engineer can point out that, from this economy limit on upward, throttle pushing will pay amazingly small dividends in speed. How little dividend it pays is best shown by two examples: If you could double the horsepower of an airplane whose top speed is 100 m.p.h. (double it, that is, without adding weight or bulk), the airplane's top speed would still only be 126 m.p.h.! If you wanted to double the top speed of an airplane that has a 65-horsepower engine, you would have to give it an engine of 520 horsepower!

MORE PHYSICS

The physical law behind these facts is easy to understand. In pushing an object through a fluid, the drag increases as the square of the speed; but the power required increases as the cube of the speed. This applies to boats, bicycles, cars, airships; and in high-speed flight it applies also to airplanes. The best way to prove the cube law for yourself is on a bicycle. You try to pedal a little faster. You notice right away that you must bear down on the pedals with a great deal more force. That is the drag of the air on your body, increasing as the square of the speed. Still, "This is all right," you think, "I have muscle enough to go even faster if I wanted to." But a minute later, you realize that you are not only bearing down much harder on the pedals but that you must of course make this greater exertion also much more often per minute; that is the cube-law effect which makes the power requirements so high; and you presently find that you do not have power enough to keep up such speed.

In the case of an airplane flying at slow or intermediate speed the

---

* Military ships, and especially military trainers, are built for much higher load factors than 6. If you are flying Army trainers, you can just about forget about your structure.

working of the cube law is mitigated somewhat by the induced drag, the drag which the wings develop as part of their lifting action. This drag—peculiar to airplanes and not found in other conveyances—is extremely high when an airplane is flown slowly; it decreases rapidly as the airplane is speeded up. At about cruising speed, it becomes of negligible importance. This means that, when a pilot is flying his ship at very low speed, and he then tries to fly a little faster, he thereby actually decreases the total drag of his airplane! And as far as power requirements are concerned he thus does not get the full penalty of the cube law: because the drag fails to increase fully as the square of the speed, the power requirements fail to increase fully as the cube of the speed. But when the pilot is already flying fairly near top speed, where induced drag has almost disappeared and parasite drag is the only important kind of drag, and if he *then* tries to fly a little faster, he does get the full working of the cube law straight in the neck.

SPEED OF FASTEST CLIMB

An airplane's rate of climb depends on its excess power, that is, on the difference between the amount of power that is necessary simply to sustain flight, and the amount of power that the engine-propeller combination is actually putting out. Whatever speed leaves you the biggest margin of excess horsepower will also give you the best rate of climb.

If you think that over, you come to the conclusion that the airplane's rate of climb must be best if it is flown at wide-open throttle, at the Speed of Best Duration, that is, only a few miles per hour faster than its stalling speed; for, as has been shown, that speed is also the Speed of Least Power Requirement. But, as every pilot knows, the airplane does not behave that way; it does not climb fastest if you point it up too steeply. The reason why it doesn't is that, as such slow flying speeds, the power plant cannot develop its full horsepower. Even though the throttle may be wide open, the engine is handicapped because it can't rev up to full r.p.m.; and the propeller is inefficient because its blades are designed to work with maximum efficiency somewhere near cruising speed. Thus, when you fly at very slow speed, you do indeed need very little power simply to sustain flight. But at the same time power output is also small, and thus the margin of excess horsepower is not impressive.

As the airplane is speeded up from near-stalling speed toward intermediate speeds, its power *requirements* increase, as we have seen, because, even though the drag decreases, the higher speed itself represents an additional demand on power. But the power *output* at wide-open throttle increases even faster—because the engine starts revving up and the propeller starts to bite better. The result is that most airplanes climb fastest when flown rather fast—at approximately the same speed that will furnish the shallowest glide.

Of all the working speeds of the airplane this is the one whose exact location on the speed scale is hardest to fix; too much depends on the exact characteristics of the power plant—r.p.m., propeller pitch and diameter, and so on. It is believed that in most airplanes the Speed of Fastest Climb is somewhat slower than the Speed of Best Distance—and that is the way it has been indicated in the diagram. But the airplane is not very sensitive to speed changes as regards rate of climb. Flying it 10 m.p.h. faster or 10 m.p.h. slower than the Speed of Fastest Climb will still give you almost the same rate of climb. In practice, a pilot should always favor the higher air speed. The brisker air flow will mean better cooling of the engine. The higher r.p.m. of the engine will mean less tendency to detonation or knocking.

THE SPEED OF STEEPEST CLIMB

An interesting problem is the steepest climb. In climbing out of an obstructed field and in possibly one or two other situations, the pilot is less concerned with his actual *rate* of climb, in feet per minute, than with his *angle* of climb—how many feet of altitude he will have when he gets to the edge of the field. It would take a good many words to present here the reasoning by which this speed is calculated; it is enough to state here as a fact that an airplane will climb the steepest if at wide-open throttle the nose is pointed well up and its speed is allowed to become comparatively quite slow.

This is what you might call a *dangerous thought;* by implication it seems to say that, after taking off, a pilot should make his initial climb steep and slow. Actually, it doesn't say that he *should*. In routine flying, it is seldom necessary to achieve the steepest possible climb; and considerations of engine wear alone should keep the pilot from climbing unnecessarily steeply, also the thought that he wants a margin of safety in case of engine failure or gustiness.

The inexperienced pilot may then actually be best off not knowing this fact at all. For it has real importance mostly in an emergency, as when climbing out over obstructions; and in such an emergency, the danger is great anyway that the inexperienced pilot will haul back on the stick in his anxiety and will stall or spin.

But the experienced pilot who can trust his own reactions might as well know this fact and make use of it. A certain type of small airplane which is reputed to be "underpowered" and sluggish on the take-off is an example. The real trouble with the ship is merely that pilots are afraid, right after the take-off, to haul back and hold it nose up as firmly as it should be held up, and as in that particular ship it may be held up with perfect safety.

Slow flight does add an element of risk. But on the other hand, if a steep climb will let you clear an obstruction by a more comfortable margin, then that is in itself an element of safety.

"DOES THIS MEAN MY SHIP?"

A pilot might well wonder whether all this applies to his particular ship. It probably does. If his ship has an unusually wide span, it will be unusually efficient in slow flight, and the best speeds both for duration and for distance as well as the best climbing speeds will move to the slow side. It will pay to fly it extra slowly.

If his ship has rather short wings, it will be inefficient in slow flight and the best speeds move to the fast side. It will pay to fly it a little faster than a little slower.

If a ship bristles with struts, wires, cylinder heads, and windshields, it will be inefficient in fast flight, and the best speeds move to the slow side again, especially so the limit of cruising economy. If it is very clean, the pilot should favor slightly higher figures.

A controllable-pitch propeller or perhaps a seaplane propeller or otherwise a propeller of flat pitch will make the power plant extra-efficient at slow speeds and would put all the working speeds a little lower; a small, fast-turning propeller would make your power plant best at higher speeds, and all speeds that involve power would go to the high side, especially so the power-on Speed of Best Distance and Best Duration.

If heavily loaded, all speeds are higher; when very light, all are lower. During a long flight, for example, your Speed of Best Distance becomes slower as you burn up fuel. Only your maneuvering speed had best remain fixed at twice the normal stalling speed for the ship's normal weight; and remember—take it easy.

Chapter 20

THIN AIR

THERE'S gold in them thar hills—but you will have to go up there to get it. The hills—that means the upper altitudes— 5,000, 10,000, 20,000 feet, according to the size and power of your ship. Every pilot knows that the big supercharged airplanes, Stratoliners, Flying Fortresses, and such, gain speed and efficiency by flying high. But many pilots don't know just exactly how those gains are made. And many pilots don't realize that the small unsupercharged airplane also becomes more efficient at high altitude. In fact, the popular belief is that the ordinary small airplane, with its fixed-pitch prop and putt-putt engine, performs best right near the ground.

This is true only in certain respects. It is true that the ordinary airplane's top speed is highest near sea level; its rate of climb is best there, the landing is slowest, and the take-off shortest. But for the most important item of performance, workaday cruising, sea level is not best; for cruising, altitude makes any airplane more efficient. The air lines found that out long ago and have long gone in for scientific control of cruising altitude and cruising power. The time has come when the "nonscheduled" flier, too, should understand the economics, as it were, of high-altitude flight.

It is not a simple story. Perhaps it will be clearest if we state first of all, without going into the reasons, just how the airplane's behavior at 10,000 feet differs from its sea-level behavior. Take an ordinary airplane with unsupercharged engine and fixed-pitch propeller. Say that at sea level it cruises exactly 100 m.p.h., at a fuel consumption of 5 gallons per hour, so that in still air it does 20 miles to the gallon. Say that the range is 4 hours.

In such an airplane you are 400 miles from home, with 5 hours of daylight left. You know the situation—how often things break just that way, and how annoying it is. You wish you had a faster airplane or one with longer range. Either would do the trick. As it is, you will

have to make an intermediate landing. It will kill an hour; and if service is as slow as usual there won't be enough daylight left to get home:
you'll have to spend the night.

All right, says the engineer; take a pencil and mark your sea-level
cruising throttle setting. Take the airplane upstairs, leaning out your
mixture properly as you go up. And then, with that same throttle
setting you used for sea-level cruising, try level flight at 10,000 feet.

The air speed will now show about 79 m.p.h.; this does not look
attractive. The airplane will fly slightly nose-high, mushing; to any
right-thinking pilot this looks bad. The engine has lost some 150 r.p.m.
The whole thing looks wrong, sounds wrong, and feels wrong.

SOMETHING FOR NOTHING

But the whole thing is rather impressively right. If you investigate,
you find that almost everything is in your favor. The 79 m.p.h. indicated by your air-speed indicator is actually, at this altitude, 93 m.p.h.;
you have not really lost so much speed. And the hourly fuel consumption at this altitude, with this throttle setting and r.p.m. and the mixture properly leaned out, is only a little more than 3 gallons, instead
of 5! This means you are getting 30 miles to the gallon instead of 20! It
means that, instead of 4 hours, you can stay up 5 hours and a half!
It means that instead of 400 miles your last-drop, still-air range is 510
miles! It means you can make it home in one hop of 4 hours and a
quarter, arriving with 40 minutes' daylight to spare and more than an
hour's fuel still in reserve!

That's what high altitude can give you: something for nothing.
This simple example disregards wind effects, and it disregards also
some unfavorable factors, such as the climb to altitude, and that you'll
feel lonesome and shivery up so high, and that the ground will seem
not to move at all, and you'll be bored. But it shows that the airplane
becomes more efficient with altitude.

In this case, you get the benefits of this improved efficiency not in
form of cruising speed but, at only a slight sacrifice in speed, in form of
range and endurance and thus in actual traveling speed. You could
take the same benefits also in different forms. For instance, you could
fly so that your air-speed indicator would read 85 m.p.h., advancing
your throttle a little as compared with sea-level cruising. Your true air
speed would then be 100 m.p.h., exactly what it was at sea level, and

still you would save fuel, stretching your 400-mile range by about 50 miles: thus you can have your cake and eat it too! Again, you could choose to fly at wide-open throttle, running your engine a little faster than sea-level cruising r.p.m., but getting an indicated air speed of 93 m.p.h. which at that altitude would mean an actual air speed of 110 m.p.h.! In this case your fuel saving would be slight, but you would still save time; and, while you would be over-revving your engine, this would be less serious at altitude, since manifold pressure would be low and the engine would not actually be putting out excess horsepower.

Thus you can take your free-for-nothing gift in any shape you like. The most efficient is the first—to leave your throttle at the usual setting, make a slight sacrifice in speed, and gain enormously in range; for in unscheduled flying—with no ground crew waiting to service you, it is range every time, rather than a few more miles per hour of air speed, that will determine your actual traveling speed; what really slows our small airplanes up is time wasted on the ground.

This emphasis on fuel economy may somehow suggest that efficient high-altitude flight must always mean one of those desperate last-drop fuel-stretching adventures that make a pilot wonder what will bring him down first, engine failure or heart failure. Of course, it means no such thing. A pilot might plan on having an hour's fuel reserve upon arrival, or even 2 hours'; he will still be interested in getting the most miles out of a tankful of fuel—not because of the gas money it saves but because of the time waste it cuts out. Obviously a pilot would not undertake such a flight without first checking by actual test his rate of consumption at various altitudes. Obviously he needs an accurate fuel gauge, or better still a definite reserve tank. But all airplanes need that anyway, and those that have no definite fuel reserve arrangement ought to have one. Obviously also, if the ship has a carburetor without mixture control, the whole thing is off; flying high without leaning the mixture means just pouring that good gas into thin air. An engine with fuel injector leans out its own mixture without help from the pilot.

A pilot need not necessarily understand the theory-of-flight logic behind all this. He can choose to take it on faith. What interests the pilot primarily is not why an airplane behaves the way it does, but how its behavior can best be used for efficient flying.

PRACTICAL PLANNING

Unless your airplane is very fast, your choice of altitude must depend on the wind. True air speed will decrease slightly with altitude, as we have seen, and it is the fuel economy that improves. But there is a rule in navigation that, when flying in a head wind, you should fly fast. If the head wind is strong, you should fly fast at almost any sacrifice of fuel economy. For in a strong head wind, you might in an extreme case be reduced to almost no ground speed at all, and in that case all the fuel economy in the world would be meaningless; you would burn up all your fuel while getting nowhere. This argues that, in a head wind, the high-altitude effect is useless for the small unsupercharged airplane. Moreover, the winds generally increase with altitude. This reinforces the argument that, in a head wind, it does not pay to fly the airplane high.

With no wind or a tail wind, the opposite is true; it pays doubly to fly high. Endurance increases with altitude; and there is a rule in navigation that, when flying in a tail wind, you gain most in the long run by flying slowly, economically, letting the wind do your work for you. Moreover, the tail wind is going to be stronger higher up. This reinforces the argument that, with a tail wind, you should go high, lean out your mixture, skip an intermediate landing that might otherwise be necessary and thus make a much better time.

HOW TO GET YOUR MONEY BACK

There is the climb to be considered. It takes, of course, a good deal of extra power, that is, extra gasoline. This may seem like waste— but most of it need not be waste. The waste of a climb lies in the fact that the propeller becomes less efficient in a climb and wastes much energy (which means fuel) in needlessly stirring up the air. The climb will therefore be least wasteful if it is made gently, gradually, at almost cruising throttle, and with the air-speed indicator reading a few miles per hour faster than what it reads in the normal glide.

Aside from propeller waste, the extra fuel consumed during the climb is not wasted. It is simply turned into altitude instead of distance; eventually, during the descent, the altitude can in turn be converted into distance. To do that, the airplane must be flown efficiently during the descent. The temptation is always to leave the throttle set, retrim the stabilizer, and dive slightly, converting altitude into soul-satisfying speed and stiffness of controls. But speed beyond a certain point is always wasteful in an airplane. Anytime—be it climbing, level flight, or descent; be it at high altitude or low—that your airspeed indicator reads more than a few miles per hour higher than it reads in your normal glide, you are wasting energy, cutting down on your miles per gallon. For best miles per gallon, the descent should be made with the indicator showing at most 5 m.p.h. more than the speed of normal glide; the engine should be throttled back accordingly. In our assumed 100 m.p.h. airplane, an efficient slow descent from 10,000 feet with the air speed indicating 80 m.p.h., compared with an inefficient shallow-diving descent at 115 m.p.h., will stretch the range by something like 50 miles!

How well a ship can take advantage of the high-altitude effect depends also on its tanks—how many hours' supply of fuel it carries. If yours is one of those enviable 8-hour affairs, you have all the endurance you need for practical flying over civilized country; and, since unsupercharged airplanes gain in range rather than speed, you won't be so interested in high-altitude flight.

If yours is the usual 3-hour tank of the standard light airplane, you may be unable to take advantage of the high altitude. In pure first-sight theory, you could gain by high-altitude flight. In pure and simple theory, the way to fly such a light plane would be to climb about 10,000 feet, using about an hour for the climb and keeping the air-speed reading slightly above the speed of the normal glide. Then cruise level a short while, still keeping the air speed slightly above the normal glide reading, and descend with the air-speed reading still the same, using perhaps about an hour for the descent. But the inefficiency of the propeller in the climb probably wastes more energy than your short stay at high altitude can gain back for you. This procedure probably won't pay unless the ship is very lightly loaded.

The man who really should be interested in high-altitude economy is the man with the usual 4- or 5-hour tank. He has enough endurance to stay at altitude long enough to make use of its good effects. At the same time his range is so short that he will be grateful for any extension.

WHAT ALTITUDE?

Finally, the correct choice of altitude is important. The most efficient altitude depends on the airplane's exact design and also on its load. For most airplanes, the most efficient altitude should be about two-thirds of the airplane's service ceiling. But there is a simple way to find out exactly where it is. You know what your particular air-speed indicator reads in a normal glide (with all its errors, however crazy). Add 5 m.p.h. or so; that is the Speed of Best Distance, power on—the flight condition in which your airplane will make the most miles per gallon. And *that* is the mark at which you want to keep your air-speed indicator. If at any given throttle setting, your air-speed indicator in level flight indicates higher, you are not yet at your best altitude for that throttle setting, you can still gain in miles per gallon by going higher. If at any given throttle setting your air-speed indicator indicates lower, you have too much altitude for that throttle setting, and you gain miles per hour by opening the throttle more or by going lower.

CAUSES AND EFFECTS

Now for the logic behind all this. A pilot needn't understand it, but he may wish to understand it; for high-altitude flying, "over-weather" flying, stratosphere flying is becoming more important every year; and the principles involved in altitude flight are the same, whether you are flying a Cub at 10,000 feet or a Lockheed Lightning at 40,000 feet.

It is a long chain of causes and effects.

The basic effect, the one on which all high-altitude flying depends, is the effect that the thin air of high altitude has on the lift and drag of an airplane, namely, that it has no effect at all! Imagine an airplane flying in thick sea-level air, cruising level. Let it be a small two-seat airplane, weighing 1,200 pounds. As it cruises, all forces on it are in balance; 1,200 pounds of weight pull down; 1,200 pounds of lift push up. A certain amount of drag, say 200 pounds, holds it back. Two hundred pounds of propeller thrust pull it forward. The result is steady level flight at 100 m.p.h. Suppose now that this airplane should suddenly burst into an area of much thinner air—high-altitude air. The lift immediately would fade out, since each cubic foot of the thinner air would have so much less mass and the wings would have less to push against. The airplane begins to mush. But the drag also fades out, for the same reason; and with 200 pounds of propeller thrust still pulling on the nose, the airplane would begin to speed up. As it speeded up, the mushing would stop. And presently, the airplane would steady down, with the same lift and the same drag as before, in the same flight condition as before, (that is, at the same Angle of Attack) but in much faster flight.

This is the key to understanding high altitude. This is how we get that something for nothing; it is, in this its simplest form, additional speed for no additional drag. And it means much more than is apparent at first glance. Take that little airplane to 36,000 feet, and it will fly twice as fast with still only the same 200 pounds of propeller pull to keep it going. If, on the other hand, you tried the same thing at sea level, that is, tried to make that same little airplane cruise twice as fast, its drag would be four times as strong; requiring four times the propeller pull to keep it going!

HOW YOUR AIR SPEED MISBEHAVES

Next in the theory of high altitude comes a minor item which is, however, of major importance to the pilot—what the thin air does to the air-speed indicator. The air-speed indicator shows true air speed only at sea level; at altitude, it reads too low. In fact, it is not really a speed indicator; what it really indicates is something else. There is a rule of thumb that, in order to get your true air speed, you must add 2 per cent to your indicator reading for every thousand feet of altitude. For example, if your indicator reads 100 m.p.h. and your altitude is 5,000 feet, your true air speed is 100, plus five times 2 per cent of 100; that adds up to 110 m.p.h. This rule of thumb holds true only at low altitudes. For accurate work, pilots use a special slide rule that takes account of altitude and temperature and converts "indicated" air speed into "true" air speed; it is by such a slide rule that you know, when flying at 10,000 feet with the air-speed indicator reading 79 m.p.h., that you are really doing 93.

The instrument may not work so well as an indicator of speed, but it works beautifully as an indicator of flight condition, of Angle of Attack. Suppose that your airplane stalls at sea level when your air-speed indicator reads 45 m.p.h. It will then stall whenever the instrument reads 45 m.p.h.—regardless of the altitude. If you stalled your airplane at 36,000 feet, where that airplane's actual stalling speed would be 90 m.p.h., the air-speed indicator would still indicate only 45 miles per hour. Thus you can make a mark, mental or actual, on the instrument's dial and label it "stall" and it will be valid for all altitudes! The same thing is true of other flight conditions, for example, the glide. If in your best glide at sea level the indicator reads 75 m.p.h., then at any altitude, your best glide will be whatever makes your air-speed indicator read 75 m.p.h. At 36,000 feet the best glide would actually be at a speed of 150 m.p.h., but on the air-speed indicator, it would show up as 75 m.p.h. just the same. Hence you can label that place on the indicator dial "best glide," and then, regardless of altitude, regardless of actual speed, you can always get the best glide by flying so that the needle will be on that mark.

Why is this so? The wing and the air-speed indicator are devices of the same sort and are subject to the same laws. The wing is a shape on which flowing air builds up a dynamic pressure which holds the airplane up. The air-speed indicator is a shape—a tube into whose open end the wind blows—on which flowing air builds up dynamic pressure which deflects the needle. That is what the instrument really measures, a dynamic pressure. If the air comes flowing against the wing heavily enough and fast enough to hold the airplane up, it also comes flowing into the Pitot tube heavily enough and fast enough to run the instrument's hand to that certain place on the dial. If the air gets too thin or the speed too slow, the pressure on both wing and Pitot tube fades: the wing mushes, and the air-speed indicator needle shows less deflection. If the speed then builds up again the pressure on both wing and Pitot tube builds up again, the wing stops mushing, and the air-speed indicator needle goes back to the old place. Thus wing and Pitot tube keep step with each other. What happens to one happens to the other.

This also gives you warning of a most disappointing fact. Few things would look nicer to a pilot than to see an air-speed indicator indicate something like 400 m.p.h. Well, you probably won't ever see it—not in level flight. The spectacular high speeds that you hear about are achieved with supercharged engines at high altitude; and the pilot who cruises some twin-engined interceptor at 400 m.p.h. at 30,000 feet sees his air-speed indicator register a measly 250!

POWER VS. FORCE

Next in the logics of high altitude, there is an engineering fact that is sometimes overlooked when pilots discuss an airplane's altitude behavior: At higher altitude, the airplane needs more power. In any given flight condition—such as flat cruising flight or slightly mushing flight or nearly stalled flight—it will fly faster at the higher altitude; it will fly faster without additional drag. But it nevertheless will need more power.

This seems contradictory; if at the higher speed there is no more drag, then why does it take more power? The answer is really just simple physics—the difference between "force" and "power." "Force" is a pull, measurable in pounds. Lift, drag, and propeller pull are forces. "Power" is the product of force and speed; it contains the elements of time and distance as well as force. Power means pulling something fast, and moving it through a distance. What the motor turns out is power. In the stratosphere, the airplane needs the same force pulling on its nose as at sea level; but, since that force must pull twice as fast, the propeller must turn twice as fast or be twice as big or have twice the pitch; and to turn that propeller takes twice as powerful an engine.

At first glance, this might seem to nullify all the advantages of high-altitude flight; to fly twice as fast with twice the power looks like no bargain. But it means more of an advantage than is apparent at first glance. Try at sea level to cruise an airplane twice as fast, and you would need not twice the power, but eight times the power! You would have to fill the airplane so full of engine (and of gasoline to run that engine) that it could not carry anything! And the beauty of stratosphere flight is that, flying twice as fast with twice the power, you get from A to B in half the time; and, since you run your twice-as-powerful engine only half as long, you use up no more fuel than you would have used for the same trip in slower flight at sea level.

Thus the something for nothing pops up again here; it now shows up as additional speed, free of charge.

But this fact—that the airplane needs more power at the higher altitude—is important also in other respects. It means that any airplane, regardless of its power plant, will gradually run out of power as you take it up high. Even if altitude did not affect the power plant and its horsepower output, this effect would nevertheless at some altitude force the ship to mush for lack of sufficient power. That's why, for real high-altitude performance, you need a wing that is shaped right for mushing flight—a gliderlike wing of high aspect ratio—narrow chord and long span. That's largely why B-17's and P-38's are shaped the way they are; at really high altitude, any airplane mushes. Still higher up any ship would stall from lack of power. Any airplane, regardless of power plant, and however supercharged, has a ceiling.

SELF-THROTTLING ENGINES

But actually, the power plant loses power as you go up. This is the most familiar of all altitude effects. You sense that loss of power even at 3,000 feet trying to get altitude for spins. In supercharged engines, this power loss may not show for the first 5,000 or 10,000 feet of altitude. But above its critical altitude, even a supercharged engine behaves just like a putt-putt.

The exact nature of this power loss is worth understanding. The pilot sees his throttle wide open and at the same time feels that he gets very little power. To him it may easily seem that this shrinkage of power represents a genuine loss, a waste of energy, a useless burning up of fuel. Actually, it does not. Actually, the main reason why the motor puts out less power is that it takes in less fuel. Efficient engine operation requires 13 pounds of air for each pound of gasoline. At the higher altitudes, each suction stroke of each cylinder cannot pump as much of the thinner air into the engine as it can get at low altitude. Hence it can burn less fuel. And either manually, by the mixture control, or else mechanically by various gadgets, the fuel input is adjusted to the air intake.

Thus the power loss at altitude is not really a loss at all. What really happens is that, in a fancy and indirect way, the engine throttles itself down.

There is of course also some genuine waste in high-altitude operation of the engine. The internal friction of the engine, for example, the friction of metal sliding on oil-covered metal, the friction of mixture flowing through narrow tubing, squeezing through narrow valves—all that remains practically the same regardless of altitude. And as the power generated in the engine diminishes, this friction becomes proportionately more and more important.

This also explains something that sometimes is heatedly argued by pilots: Will an engine rev up with altitude, or keep constant revs, or rev down? If it were not for that factor of internal friction, an engine and propeller would keep constant revs regardless of altitude; for the same thinning out of air that reduces engine power would reduce propeller resistance to exactly the same degree. But, because the engine suffers from this internal friction, while the propeller is free of any corresponding effect, the unsupercharged engine with fixed-pitch propeller will actually rev down as you go up. At 70,000 feet or thereabouts, a gasoline engine, however supercharged, will turn out just enough power to overcome its own internal friction and keep its superchargers going—but not enough to turn any propeller.

Then there may be an additional and serious waste of energy in the propeller itself. Designed for one combination of air speed and r.p.m., the propeller goes aerodynamically out of whack and works wastefully at any different combination. This is largely why the stratosphere airplane simply could not fly without a controllable-pitch propeller. For the unsupercharged airplane in cruising flight at 10,000 feet or so, the propeller waste is not too serious. Altogether, then, it still is true that when you fly at high altitude you are simply throttled well back, getting less power, and using up less fuel.

Thus high altitude squeezes the airplane from two sides. With one hand it forces it to fly faster to keep from mushing or from stalling. With the other hand it throttles the motor down. The result is that the airplane is forced to mush; it is forced to fly at higher Angle of Attack in exactly the fashion in which it flies near sea level when the pilot maintains level flight with very little power.

TAKE IT EASY IN A HURRY

And this is where the logic of high-altitude flight finally clicks to a happy conclusion. To most pilots, this sort of slightly nose-high flight, "mushing," is an abomination. Actually it is the most efficient way for an airplane to fly! Regardless of altitude, any airplane will always fly with the least waste of energy, that is, will go the most miles to the gallon, if it is flown with the nose well up, the engine throttled well back, and the air speed reading only a few miles per hour faster than what it reads in the normal glide. What we call "cruising" is from the engineering point of view barbaric waste. If we fly a Cub faster than

55, or a Culver Cadet faster than 90, or a B-26 faster than perhaps 200, we do so not because that is the way to fly them. We do it only because we are so darned impatient.

At sea level, then, if you wanted to get the most miles per gallon out of an airplane that "cruises" at 100 m.p.h. and stalls at 50, you would fly it at 80 m.p.h. slightly mushing; but 80 m.p.h. would be too slow for your impatient soul, and hence you wouldn't do it. At 10,000 feet, it is still true that this slow slightly nose-high way is the most efficient way to fly. But at that altitude, with an engine that has throttled itself back, there is no choice about it; it is practically the only way in which we can fly. And now at 10,000 feet, this whole condition occurs when the airplane is actually moving through the air at 92 m.p.h., which should be almost fast enough for the impatient pilot.

That is the secret of high-altitude flight; you can have your cake and eat it, too. You can mush at high speed. You can take it easy aerodynamically and yet move in a hurry across the map.

REMEMBER, NOW:

As a pilot, you need not understand all this as long as you know the final effects—that at altitude the airplane becomes more efficient and goes farther and or faster on less fuel. But as a pilot, you have to understand one thing: Going up, you lean out your mixture; coming down, you have got to rich it up again; because, if you don't, the engine will quit.

MILE-HIGH AIRPORTS

But now: How does all this work out when the altitude is right on the ground? Out West, where men are men and aeronautical maps are a solid brown, a pilot must not only do his cruising at altitude willy-nilly, but he must also make approaches and land, take-off and climb out over obstructions while subject to the laws of thin air. Out there, with visibility perhaps 60 miles, the air not merely is thin but it even looks thin; you can almost see that there is little substance to it. Think of those high fields—Cheyenne, Wyoming; Butte, Montana; Grand Canyon! Or, it used to be that, flying a Cub along the Cheyenne-Salt Lake Airway, you had to stop for gas at Dubois, Wyoming, a Federal auxiliary field at elevation 7,200 feet. Or, if that doesn't seem enough of a problem, ask some pilot about Rock Springs, Wyoming, high and obstructed by bluffs. Or about Lordsburg, New Mexico, high and hot and on a slope with soft gravelly surface. Or, if you want really fancy mental hazards, think of those landing strips the Forest Service uses in Montana, Idaho, and California, short rough clearings at elevations up to 9,000 feet!

At first, it scares you. Moreover, one is likely to pick up all sorts of scary ideas—how at such elevations the airplane will glide in "like a brick"; how it will stall at much lower Angle of Attack than at sea level, or with the stick much less far back; how you had better keep some power on during the approach; and how you better keep an extra 10 or 20 m.p.h. on the air-speed indicator; and so forth. And, because of both, the actual effect of thin air and the mental hazards, one's flying the first few days out in that country may be a bit messy.

To sort this out, let's set the scene at some airport in the West. Let the field elevation be 6,000 feet, and the time a summer afternoon when the air temperature near the ground is around 100°F. For that is the combination you have to contend with in summer in the West— thin air, thinned out still more by heat. Let the airplane be any small training ship or commercial ship—in short, a simple airplane of average design, with unsupercharged engine and fixed-pitch prop.

DON'T DO ANYTHING

As for the approach to such a field, the most remarkable thing about it is that you won't notice any difference; you won't, that is, un-less the ideas you have picked up mislead you into some such fault as gliding especially steeply, or coming in high, or with power on, or otherwise trying to compensate for an effect which simply isn't there. High altitude causes no important changes in an airplane's glide. The gliding attitude is the same; the ship does not need to be pointed down any more steeply than usual. The air-speed indicator reads the same; if in a normal glide near sea level your air-speed indicator read 75 m.p.h., it will read 75 m.p.h. in a normal glide also at 6,000 feet, or at 26,000 feet! The sounds of the glide are practically the same. The prop windmills at practically the same r.p.m. Most important of all, the flight path is the same; if near sea level you could glide 8,000 feet of distance from 1,000 feet of altitude, you will do the same also at 6,000 feet.

The only difference that altitude makes in your glide is that your actual air speed is faster. The airplane glides along the same flight path as at sea level, and with the same air-speed-indicator reading, but it glides along that flight path at a faster actual speed. At 6,000 feet you are gliding about 10 per cent faster because the air at that altitude is thinner; and you are gliding another 5 per cent faster because the heat has further thinned out the air. It might be a good rule to remember that on the average Western airport in summer, your actual air speed is about 15 per cent higher than at sea level. But then again, it might be even better to remember no rule at all, but simply to disregard the high altitude and fly as usual. If you try to use any special technique, it will only mess you up.

Student pilots often find it hard to believe this. One might think that in the thinner air he must glide with a higher reading on the air-speed indicator. But here again: If the air flows heavily enough into the Pitot tube of your air-speed indicator to push the needle to the usual place, then it also flows heavily enough against your wings to give you the usual amount of lift.

Or, one might reason, if the actual air speed is 15 per cent higher, then surely it must be necessary to nose down the ship more to get that higher air speed: but here again the airplane gives itself that higher air speed without the help of a steeper glide—simply because the air is thinner. Suppose you were gliding along in ordinary air at a normal glide, and you suddenly entered a region of much thinner air, the ship would very quickly adjust its own speed. In the thinner air it would have less lift; it would therefore begin to mush. But at the same time it will have less drag; and, since gravity would keep pulling it along with undiminished force, it would speed up! As it would speed up the mushing would stop. And presently, the ship would be steadied down in a normal glide, at the same angle as before, but at a higher speed.

The actual landing, too, is remarkable mostly for the things that don't happen; it is quite similar to your usual landing at a sea-level field. The ship will float and settle in its usual sea-level manner. It will stall when the air-speed indicator shows your usual sea-level stalling speed. In theory, the thinner air makes the ship a little livelier, "hotter"; its responses less well-damped, less Cub-like, more jittery. But you would have to go to 20,000 feet or so before that effect would be strong enough to notice. The only real difference is again merely the actual speed. The actual speed will again be 10 per cent higher because of the altitude and an additional 5 per cent higher because of the heat. And in the landing, that does become noticeable.

According to the mechanical law that governs stopping distances in automobiles and landing runs in airplanes, this 15 per cent higher speed makes your landing run at the average Western airport about one-third longer than at a sea-level airport. It makes any bouncing, any roughness of ground, any jolt about one-third harder, increasing the stresses on the ship by one-third. It also noticeably increases the ship's tendency to ground-loop, requiring more alertness on the rudder. But even so, all that requires no special technique. If your usual technique is any good, it will see you through also at 6,000 feet or even at 16,000 feet. Any attempt to use any special technique will here again only mess you up.

SOMETHING WRONG WITH THAT ENGINE

The take-off from that same field will be a different story—a sad one. The take-off run will be very long indeed, the climb much slower than at sea level; what's worse, the angle of climb will be much shallower. Altogether, the first time it happens to you, you won't quite believe it, especially not the way the engine has somehow lost its "stuff" and how your hand on the throttle fails to get any real results— all noise and no pull, it seems. Pilots have been known to return to the hangar and declare, "Something is wrong with that engine."

This effect is due to three causes—the need of the airplane for more power, due to higher speed; the self-throttling of the engine with altitude; and the power waste in the propeller—which is especially serious in the take-off and climb. Altogether, if the altitude is 6,000 feet and the temperature 100 degrees, your airplane will take off and climb about as it would at sea level, if—in a ship that cruises at 2,250 r.p.m.—you tried a take-off with perhaps 1,600 r.p.m. Try it sometime! It will teach you not to load your ship to the limit for your first trip West.

Just how much room you will need for take-off is impossible to calculate since so much depends on small factors. Much depends, for example, on the type of runway—tall grass has fifteen times as much drag on the wheels as has concrete—and on any upslope that may hinder you or downslope that may help you. At sea level, when the ship has lots of excess power, those things don't matter much; at altitude they become quite disproportionately important. Altogether, in an average ship under average conditions, your take-off run at 6,000 feet that hot afternoon is likely to be two or three times as long as at sea level; but, if the ship is heavily enough loaded, the engine worn enough, and the air hot enough you may get an infinite run: you simply fail to get off.

Your initial climb, too, is going to be poor and scary; near the ground, where the air is hot, the rate of climb will be only about half the sea-level rate; at the same time your forward speed is higher, and thus your angle of climb will be less than half your sea-level angle; even quite distant obstructions must be taken seriously.

But perhaps most disconcerting of all are the effects of updrafts and downdrafts. Even in sea-level country, updrafts of 400 feet per minute and downdrafts of 200 feet per minute are frequently met on any day with unstable air. In the high country of the West, where the action of the sun is stronger, updrafts and downdrafts are often much livelier— and the mountains help. Even the presence of the airport itself helps; there is quite likely an updraft over the field itself, and compensating downdrafts in the vicinity of the field. With your rate of climb so low, this means that soon after take-off you may find yourself in a downdraft that you can't outclimb. With throttle wide open and the nose pointed as high as you dare, you may still lose altitude. This makes a powerful temptation to stall yourself, especially if it catches you by surprise. The thing to do is to grin and bear it and fly straight ahead in hope of better air farther on; to make a turn would almost always be the wrong policy, since a turn takes additional power and costs additional altitude.

LOCAL ADVICE NO GOOD

Western airports are of course designed with all this in mind and have long runways and clear approaches, or rather exits. But even so, you may sometimes have to make difficult decisions; whether to take off at all, and if so, which way, and with how much load. Local advice in such matters, though freely tendered, is usually no good; certainly not unless the adviser is an experienced pilot familiar with your type of ship. Nor does it matter what some other fellow (whose ship was also painted red) did at the same field 3 months ago; unless you know his ship, his load, and the weather at the time, it has no bearing on the problem. Therefore it may be worth while to do some advance thinking on such problems.

Often the choice is between taking off uphill up-wind and taking off downhill down-wind. It can be an extremely difficult choice. A take-off uphill may not get you off, but it has the advantage that it is easy to stop should you find you can't get off; the wind, the brakes and the slope will all help each other in bringing you to a stop. A downhill take-off is, of course, much easier; but, should it fail to materialize, it will be hard to stop without nosing over; the wind, the brakes, and the slope will all help each other in nosing you over. Since the uphill up-wind run can be tried without serious risk, it may be worth trying first in any case. If, on the other hand, the uphill run leads toward any sort of obstructions, it may be hopeless, since after lifting off you will have to climb anyway merely to stay off the runway and may be unable to get the additional climb necessary to clear the obstruction. Because of the perspective, one probably tends to underestimate the actual climb that is necessary to clear an obstruction beyond the end of a rising runway.

The one decision that is almost certainly wrong is to lift off uphill up-wind and then make a 180-degree turn to fly away downhill down-wind. The turn will use up too much power and probably altitude.

High-field experts, such as the pilots flying for the Forest Service, seem to favor the downhill take-off almost regardless of wind. In fact, many of the forest service fields are so hemmed in by mountain sides that access is possible only from one side, and the take-off has to be made through the same gap through which you came in. But the fact of the matter is that no rule can be made.

When in doubt, there are two things you can do. You can unload your ship, try some take-offs empty, and see how they go. Or you can wait for better air; and those Western pilots, the same ones that sometimes choose the downhill down-wind take-off, will sometimes wait for days before going into or out of a difficult field. Usually, a wait until the next morning is enough. In the Rocky Mountain region in summer, the early morning is about 50°F. cooler than the early afternoon. Because heat thins out the air and cold makes it denser, a high temperature thus has the same effect on an airplane's performance as if the airport were changing altitude—it cuts down the power and at the same time raises the ship's power requirements. On a hot afternoon the typical Western airport is in effect some 2,000 to 5,000 feet higher up than in the cool of the morning. At 100°F., a 6,000-foot-high airport really lies at 10,000 feet, as far as density of the air is concerned! And right at the ground, in those lowest 5 feet where the engine gets its intake air and the wings must do the take-off job, afternoon air temperature may well be 120! In short: Wait for the cool of the morning.

Once away from the airport, you may have a stiff long climb to clear a mountain range. Climbing an airplane at the higher altitudes is different from climbing the same airplane at lower altitude; it requires a different technique. Near sea level, the airplane has two best climbing speeds; the fastest climb occurs in the average airplane with average propeller, when the air-speed indicator shows about halfway between cruising speed and stalling speed. The steepest climb occurs when the air-speed indicator hovers quite near stalling speed. At altitude the speed of fastest climb moves toward the speed of steepest climb, until at the ship's ceiling they are identical—and the climb zero. Near the ceiling, the only way to squeeze any further climb out of the airplane is to fly it at the speed of steepest climb, quite near the stall with the air-speed needle only 5 or 10 miles above stalling. That is what ceiling really means; you fly in a condition where any further attempt at steeper climbing leads to a stall. Any attempt at shallower climbing leads to loss of altitude. But, even at the altitude with which we are concerned here—5,000 to 10,000 feet, the effect is noticeable; the best rate of climb is achieved if the ship is flown so that the air-speed indicator reads lower than it does for best rate of climb near sea level. The higher the altitude at which you are climbing, in short, the more firmly should you hold the ship's nose up and the lower should your air-speed indicator read.

In actual practice, the most important thing about your climb is in that high country again perhaps not what the ship itself can do but what updrafts and downdrafts do to the ship—and what you do in regard to updrafts and downdrafts. In fighting for altitude you are tempted to climb hard when you notice a downdraft and perhaps to take it a little easier when an updraft makes the altimeter creep on satisfactorily. This is a natural reaction; but, as is so often the case in flying, the natural reaction is wrong. The soaring pilot who gains altitude without any power does exactly the contrary; the technique of soaring consists in spending more time in updrafts than in downdrafts. By climbing hard in a downdraft, you slow the ship up and hence you will stay in the downdraft longer. By climbing less hard in the updraft, you allow the ship to speed up, and hence you will leave the updraft sooner. The thing to do is to use an updraft for extra steep climbing, perhaps even circling back into it to make it work for you a little longer. In a downdraft, the thing to do is to let the nose down a little, speed up, and get out of it.

How can you tell you are in an updraft? A sensitive altimeter or a quick-acting rate-of-climb indicator show updrafts and downdrafts well enough, but if you don't have those you will soon find that the quickest acting climb indicator is your own ear—the change in the sounds of the ship as the updrafts and downdrafts change your altitude. These changes are due not to any change of air flow or engine speed, but to the changing pressure in your own ear. When the sounds of the airplane become muffled and soft, you are going down. When the sounds of the airplane become loud and clear, you are going up.

And may the sounds in your ear, brave reader, be loud and clear.

INDEX